SpringerBriefs in Business

Organisational Studies

Series Editor

Jan Jonker, Nijmegen, The Netherlands

For further volumes:
http://www.springer.com/series/11458

Wayne Visser

CSR 2.0

Transforming Corporate Sustainability and Responsibility

Springer

Wayne Visser
Kaleidoscope Futures Lab. Ltd.
London
UK

ISSN 2191-5482 ISSN 2191-5490 (electronic)
ISBN 978-3-642-40873-1 ISBN 978-3-642-40874-8 (eBook)
DOI 10.1007/978-3-642-40874-8
Springer Heidelberg New York Dordrecht London

Library of Congress Control Number: 2013950022

Printed on acid-free paper

Springer is part of Springer Science+Business Media (www.springer.com)

Contents

1 Introduction 1
1.1 A Critical Perspective 1
1.2 Defining the DNA of CSR 3
References 5

2 The Stages of CSR 7
2.1 Defensive CSR 7
2.2 Charitable CSR 10
2.3 Promotional CSR 12
2.4 Strategic CSR 14
2.5 Transformative CSR 16
References 18

3 The Failure of CSR 1.0 21
3.1 Ecosystems Decline 21
3.2 Climate Change 22
3.3 Poverty 24
3.4 Inequality 25
3.5 Corruption 26
3.6 The Limitations of CSR 1.0 27
3.7 Peripheral CSR 28
3.8 Incremental CSR 30
3.9 Uneconomic CSR 31
References 33

4 CSR 2.0 as a New Metaphor 35
4.1 Introduction to Web 2.0 35
4.2 Application to CSR 2.0 36
4.3 Long Tail of CSR 2.0 37
4.4 Principles of CSR 2.0 39
4.5 Shifting from CSR 1.0 to CSR 2.0 40
4.6 Web 2.0 as a Transformational Tool 42
4.6.1 Principle 1: Openness 42

4.6.2 Principle 2: Peering 44
4.6.3 Principle 3: Sharing 45
4.6.4 Principle 4: Acting Globally 46
4.6.5 Principle 5: Mass Customization 47
4.7 Conclusion 48
References 48

5 The Principles of CSR 2.0 51
5.1 Creativity 51
5.2 Scalability 55
5.3 Responsiveness 57
5.4 Glocality 58
5.5 Circularity 61
References 63

6 Assessing CSR 2.0 Performance 67
6.1 Stages of CSR 68
6.1.1 Defensive CSR 69
6.1.2 Charitable CSR 69
6.1.3 Promotional CSR 70
6.1.4 Strategic CSR 71
6.1.5 Transformative CSR (or CSR 2.0) 72
6.2 Principles of CSR 2.0 73
6.2.1 Creativity 74
6.2.2 Scalability 75
6.2.3 Responsiveness 75
6.2.4 Glocality 76
6.2.5 Circularity 76
6.3 DNA of CSR 2.0 77
6.3.1 Value Creation 78
6.3.2 Good Governance 79
6.3.3 Societal Contribution 80
6.3.4 Environmental Integrity 80
6.4 Conclusion 80
References 81

7 Conclusion 83
7.1 Implementation of CSR 2.0 83
7.1.1 Challenges and Responses 83
7.1.2 Taking Action 85
7.2 Towards Responsible Capitalism 86
References 87

About the Author 89

Figures

Fig. 5.1 CSR pyramid for developing countries 59
Fig. 6.1 Stages of CSR (illustrative scores) 68
Fig. 6.2 Principles of CSR 2.0 (illustrative scores) 74
Fig. 6.3 DNA of CSR 2.0 (illustrative scores) 78

Tables

Table 1.1	DNA of CSR 2.0	3
Table 2.1	The stages of CSR	8
Table 3.1	The limitations of CSR 1.0	29
Table 4.1	Similarities between Web 1.0 and CSR 1.0	37
Table 4.2	Similarities between Web 2.0 and CSR 2.0	38
Table 4.3	Meta-level ontological shifts	41
Table 4.4	Micro-level methodological shifts	42
Table 4.5	Web 2.0 practices for CSR 2.0	43
Table 5.1	Principles of CSR 2.0	52
Table 5.2	Local drivers of CSR	60
Table 5.3	Global drivers of CSR	61
Table 6.1	The stages of CSR	68
Table 6.2	Defensive CSR (illustrative scores)	69
Table 6.3	Charitable CSR (illustrative scores)	70
Table 6.4	Promotional CSR (illustrative scores)	71
Table 6.5	Strategic CSR (illustrative scores)	72
Table 6.6	Transformative CSR (illustrative scores)	72
Table 6.7	Principles of CSR 2.0	73
Table 6.8	Creativity (illustrative scores)	74
Table 6.9	Scalability (illustrative scores)	75
Table 6.10	Responsiveness (illustrative scores)	76
Table 6.11	Glocality (illustrative scores)	76
Table 6.12	Circularity (illustrative scores)	77
Table 6.13	DNA of CSR 2.0	77
Table 6.14	Value creation (illustrative scores)	78
Table 6.15	Good governance (illustrative scores)	79
Table 6.16	Societal contribution (illustrative scores)	79
Table 6.17	Environmental integrity (illustrative scores)	80

Abstract

This book maps the past, present and future of corporate sustainability and responsibility, or CSR. It begins by examining the evolution and current state of CSR, using a five-stage maturity model: defensive, charitable, promotional, strategic and transformative CSR. The first four stages are dubbed CSR 1.0 and characterise most current CSR practice, while the fifth stage is named CSR 2.0 (also transformative or systemic CSR) and describes emergent and future CSR practices. The metaphor of 1.0 and 2.0 is explained as an appropriate analogy for the changes needed in CSR, drawing parallels with the evolution of Web 1.0 to Web 2.0. The author argues that CSR 1.0 approaches have failed to have any significant impact on the most serious global social, environmental and ethical challenges. Three reasons for this failure are examined, namely the incremental, peripheral and uneconomic nature of CSR 1.0 approaches. The book goes on to explore the emergent CSR 2.0 in detail by elaborating on five principles underlying this new approach, including: creativity, scalability, responsiveness, glocality and circularity. Each principle is explained conceptually, as well as illustrated with progressive case studies. To complement the principles, the author introduces a four-part DNA Model, covering value creation, good governance, societal contribution and ecological integrity, which provides the basis for defining and measuring CSR 2.0. Finally, a 70-question self-assessment diagnostic tool is presented to quantify the stages of CSR, the principles of CSR 2.0 and the DNA of CSR 2.0. Sample data is used to show how the tool can be employed for future research and practitioner application.

Keywords Corporate social responsibility · CSR · CSR 2.0 · Sustainable development · Corporate citizenship · Business ethics · Responsible enterprise · Sustainable business · Triple bottom line

Chapter 1
Introduction

Abstract The concept of corporate social responsibility (CSR) is introduced and critical perspectives are shared. The limitations of CSR are discussed and its ability to significantly reduce the social and environmental impacts of business is questioned. A CSR definition is elaborated, covering what are called the four 'DNA responsibility bases' of value creation, good governance, societal contribution and environmental integrity.

Keywords Corporate social responsibility · Sustainable development · Corporate citizenship · Business ethics · Responsible business · Sustainable business · Triple bottom line

1.1 A Critical Perspective

Let me begin by declaring that, after 20 years of working as a CSR practitioner and academic, I remain a CSR sceptic. By this I mean that I am not convinced that CSR—which I define as the way in which business consistently creates shared value in society through economic development, good governance, stakeholder responsiveness and environmental improvement—has been effective. In short, if CSR is viewed as a strategy for remedying the negative impacts of economic activity, it has (so far) failed.

This may sound melodramatic, but the logic is simple and compelling. A doctor judges his/her success by whether the patient is getting better (healthier) or worse (less healthy). Similarly, we should judge the success of CSR by whether our communities and ecosystems are getting better or worse. It is true that, at the micro level—in terms of specific CSR projects and practices—we can show many improvements. However, at the macro level almost every indicator of our social, environmental and ethical health is in decline.

I will back this claim up with data in Chap. 3 on 'The Failure of CSR 1.0'. For now, suffice to say that I am not alone in my critical assessment (Blowfield and

W. Visser, *CSR 2.0*, SpringerBriefs in Organisational Studies,
DOI: 10.1007/978-3-642-40874-8_1,

Murray 2011). Nearly 20 years ago, Paul Hawken (1994) stated in *The Ecology of Commerce* that 'If every company on the planet were to adopt the best environmental practice of the "leading" companies, the world would still be moving toward sure degradation and collapse.' Unfortunately, I am convinced that this is still true today.

Jeffrey Hollender, co-founder and former CEO of Seventh Generation, agrees, saying: 'I believe that the vast majority of companies fail to be "good" corporate citizens, Seventh Generation included. Most sustainability and corporate responsibility programs are about being less bad rather than good. They are about selective and compartmentalized "programs" rather than holistic and systemic change' (Hollender and Breen 2010).

Other critics include Christian Aid (2004), which issued a report called 'Behind the Mask: The Real Face of CSR', in which they argue that 'CSR is a completely inadequate response to the sometimes devastating impact that multinational companies can have in an ever-more globalized world—and it is actually used to mask that impact.' More recently, Karnani (2010) wrote in the *Wall Street Journal* about 'The Case Against Corporate Social Responsibility', in which he claims that 'the idea that companies have a responsibility to act in the public interest and will profit from doing so is fundamentally flawed'.

Porter and Kramer (2011) admit that 'In recent years business increasingly has been viewed as a major cause of social, environmental, and economic problems. Companies are widely perceived to be prospering at the expense of the broader community. Even worse, the more business has begun to embrace corporate responsibility, the more it has been blamed for society's failures.' Porter and Kramer see CSR as something of a red herring, despite the authors having written a previous paper extolling CSR as a route to competitiveness (Porter and Kramer 2006). Now, in contrast, they characterise CSR as a 'mind-set in which societal issues are at the periphery, not the core' and 'a reaction to external pressure—[which has] emerged largely to improve firms' reputations.'

This is not the place to deconstruct these polemics. Suffice to say that they raise some of the same concerns I have—especially about the limits of voluntary action and the 'misdirection' that CSR sometimes represents. But I also disagree with many of their propositions—such as the notion that CSR is always a deliberate strategy to mislead, or that government regulation is the only solution to social and environmental problems.

There are a number of ways to respond to my bold assertion that CSR has failed. One is to disagree with the facts and to suggest that things are getting better, not worse, as do the likes of Bjørn Lomborg (2001) in his *Skeptical Environmentalist*. However, I find the evidence, which is widely available from credible sources like the United Nations, the World Resources Institute and the World Bank, both compelling and convincing.

Second, you might argue that solving these complex social, environmental and ethical problems is not the mandate of CSR, nor within its capacity to achieve. My response is that while business certainly cannot tackle our global challenges alone, unless CSR is actually about *solving* the problems and *reversing* the negative trends, what is the point? CSR then becomes little more than an altruistic conscience-easer at best; or a manipulative image-management tool at worst.

My approach is to say that while CSR as it has been practised in the past has failed, that does not mean that a different kind of CSR—one which addresses its limitations and reforms its nature—is necessarily destined to fail in the future. I call this new approach 'transformative CSR', or 'systemic CSR' or 'CSR 2.0', and its exploration is the main subject of this monograph.

1.2 Defining the DNA of CSR

Before describing the journey to CSR 2.0 in more detail, let me elaborate on the CSR definition that I began with, which forms the basis of what I call the DNA model of CSR 2.0, summarised in Table 1.1. The four DNA responsibility bases are analogous to the four nitrogenous bases of biological DNA (adenine, cytosine, guanine, and thymine), sometimes abbreviated to the four-letters GCTA (which was the inspiration for the 1997 science fiction film GATTACA). In the case of CSR 2.0, the DNA responsibility bases are value creation, good governance, societal contribution and environmental integrity.

Table 1.1 DNA of CSR 2.0

DNA bases	Strategic goals	Example indicators	Description
Value creation	Economic development	Capital investment	Economic, social, human and natural capital
		Beneficial products	Sustainable and responsible goods and services
		Inclusive business	Wealth distribution, bottom of the pyramid markets
Good governance	Institutional effectiveness	Leadership	Strategic commitment to sustainability and responsibility
		Transparency	Sustainability and responsibility reporting, government payments
		Ethical practices	Bribery and corruption prevention, values in business
Societal contribution	Stakeholder orientation	Philanthropy	Charitable donations, provision of public goods and services
		Fair labour practices	Working conditions, employee rights, health and safety
		Supply chain integrity	SME empowerment, labour and environmental standards
Environmental integrity	Sustainable ecosystems	Ecosystem protection	Biodiversity conservation & ecosystem restoration
		Renewable resources	Tackling climate change, renewable energy and materials
		Zero waste production	Cradle-to-cradle processes, waste elimination

Hence, if we look at *Value Creation*, it is clear we are talking about more than financial profitability. The goal is economic development, which means not only contributing to the enrichment of shareholders and executives, but improving the economic context in which a company operates, including investing in infrastructure, creating jobs, providing skills development and so on. There can be any number of KPIs, but I want to highlight two that I believe are essential: beneficial products and inclusive business (Hall et al. 2012). Does the company's products and services really improve our quality of life, or do they cause harm or add to the low-quality junk of what Charles Handy (2008) calls the 'chindogu society'. And how are the economic benefits shared? Does wealth trickle up or down; are employees, SMEs in the supply chain and poor communities genuinely empowered?

Good governance is another area that is not new, but in my view has failed to be properly recognised or integrated in CSR circles (Kock et al. 2012). The goal of institutional effectiveness is as important as more lofty social and environmental ideals. After all, if the institution fails, or is not transparent and fair, this undermines everything else that CSR is trying to accomplish. Trends in reporting, but also other forms of transparency like social media and brand- or product-linked public databases of CSR performance, will be increasingly important indicators of success, alongside embedding ethical conduct in the culture of companies. Tools like GoodGuide, and Covalence's EthicalQuote ranking will become more prevalent.

Societal contribution is an area that CSR is traditionally more used to addressing, with its goal of stakeholder orientation. This gives philanthropy its rightful place in CSR—as one tile in a larger mosaic—while also providing a spotlight for the importance of fair labour practices. It is simply unacceptable that there are more people in slavery today than there were before it was officially abolished in the 1800s, just as regular exposures of high-brand companies for the use of child-labour are despicable. This area of stakeholder engagement, community participation and supply chain integrity remains one of the most vexing and critical elements of CSR (Tang and Tang 2012).

Finally, *environmental integrity* sets the bar way higher than minimising damage and rather aims at maintaining and improving ecosystem sustainability. The KPIs give some sense of the ambition required here—100 % renewable energy and zero waste. We cannot continue the same practices that have, according to WWF's Living Planet Index, caused us to lose a third of the biodiversity on the planet since they began monitoring 1970. Nor can we continue to gamble with the prospect of dangerous—and perhaps catastrophic and irreversible—climate change.

A final introductory point to make is that CSR 2.0—standing for corporate sustainability and responsibility—also proposes a new interpretation for these terms. Like two intertwined strands of DNA, sustainability and responsibility can be thought of as different, yet complementary elements of CSR. Hence, sustainability can be conceived as the destination—the challenges, vision, strategy and goals, i.e. what we are aiming for—while responsibility is more about the journey—our solutions, responses, management and actions, i.e. how we get there. The challenge now is to admit that CSR 1.0 has failed, and to make CSR 2.0—weaving the strands of sustainability and responsibility—into the new DNA of business.

In the sections that follow, I start by presenting a five-stage maturity model to describe the past, current and envisaged future state of CSR. Then, I provide an explanation of CSR 1.0 and CSR 2.0 as a metaphor to describe current and emerging practice. I follow with an examination of the reasons why CSR 1.0 is failing to adequately address our global social, environmental and ethical challenges, and proceed with an elaboration of the five principles of CSR 2.0, using cases to illustrate innovative CSR approaches. Finally, I present preliminary research findings based the administration of a CSR 2.0 self-assessment diagnostic tool and conclude with ways to define, measure and implement CSR 2.0 using a four-part DNA model.

References

Blowfield M, Murray A (2011) Corporate responsibility: a critical introduction. OUP, Oxford
Christian Aid (2004) Behind the mask: the real face of CSR. Christian Aid, London
Hall J, Matos S, Sheehan L, Silvestre B (2012) Entrepreneurship and innovation at the base of the pyramid: a recipe for inclusive growth or social exclusion? J Manage Stud 49(4):785–812
Handy C (2008) Interview with Charles Handy, conducted by Wayne Visser as part of the university of Cambridge top 50 sustainability books project, 3 Sept 2008
Hawken P (1994) The ecology of commerce: a declaration of sustainability. HarperBusiness, New York
Hollender J, Breen B (2010) The responsibility revolution: how the next generation of businesses will win. Jossey-Bass, San Francisco
Karnani A (2010) The case against corporate social responsibility. Wall Street J 256(45):R1–R4
Kock CJ, Santaló J, Diestre L (2012) Corporate governance and the environment: what type of governance creates greener companies? J Manage Stud 49(3):492–514
Lomborg B (2001) Skeptical environmentalist: measuring the real state of the world. Cambridge University Press, Cambridge
Porter ME, Kramer MR (2006) Strategy and society: the link between competitive advantage and corporate social responsibility. Harvard Bus Rev 84(12):78–92
Porter ME, Kramer MR (2011) Creating shared value. Harvard Bus Rev 89(1/2):62–77
Tang Z, Tang J (2012) Stakeholder–firm power difference, stakeholders' CSR orientation, and SMEs' environmental performance in China. J Bus Ventur 27(4):436–455

Chapter 2
The Stages of CSR

Abstract The evolution of business responsibility is described in terms five overlapping ages—the ages of greed, philanthropy, marketing, management and responsibility. An 'age' can be understood as a prevailing culture or context. Each of these ages typically manifests a different stage of CSR, namely defensive, charitable, promotional, strategic and transformative CSR respectively.

Keywords Risk management · Charity · Philanthropy · Marketing · Greenwash · Public relations · Strategy · Codes · Standards · Transformation

My research and practical experience of working with companies on CSR in over 60 countries (Visser 2012a) has led me to the conclusion that the evolution of business responsibility typically falls into five overlapping ages—the ages of greed, philanthropy, marketing, management and responsibility. An 'age' can be understood as a prevailing culture or context (Visser 2011). I believe that each of these ages typically manifests a different stage of CSR, namely defensive, charitable, promotional, strategic and transformative CSR respectively.

Similar to other stage models of CSR (Zadek 2004), my contention is that companies tend to move through these ages and stages sequentially (although they may have activities in several ages and stages at once), and that we should be encouraging business to make the transition to transformative CSR in an emerging age of responsibility. If companies remain stuck in any of the first four stages, I do not believe we will turn the tide on the environmental, social and ethical crises that we face. Simply put, CSR will continue to fail (Table. 2.1).

2.1 Defensive CSR

Although greed has always been with us, I believe the modern age of greed began when the first financial derivatives were traded on the Chicago Mercantile Exchange in 1972. This marks the beginning of a trend of financial deregulation

W. Visser, *CSR 2.0*, SpringerBriefs in Organisational Studies,
DOI: 10.1007/978-3-642-40874-8_2,

Table 2.1 The stages of CSR

Dominant paradigm	Stage of CSR	Modus operandi	Key enabler	Stakeholder target
Greed	Defensive	Ad hoc interventions	Investments	Shareholders, government and employees
Philanthropy	Charitable	Charitable programmes	Projects	Communities
Marketing	Promotional	Public relations	Media	General public
Management	Strategic	Management systems	Codes	Shareholders and NGOs/CSOs
Responsibility	Systemic	Business models	Products	Regulators and customers

and the growth of the 'casino economy', which ultimately led to the boom-and-bust global financial crisis in 2008 (Visser 2010).

The Age of Greed is characterised by the ideology that 'bigger is better' and that the 'invisible hand' of the market always operates in society's best interest. In practice, however, the incentives in the market—like Wall Street profits and traders' bonuses—become perverse, leading not only to unbelievable wealth in the hands of a few speculators, but ultimately to worldwide economic catastrophe. Although the greed bubble eventually popped in 2008 with Lehman Brothers' collapse, the bankruptcy of Enron in 2001 should have taught us all we needed to know about the consequences of unregulated markets and greedy executives (Visser 2011).

In 2000, Enron had revenues of $111 billion and employed over 20,000 staff. It was named 'America's Most Innovative Company' by *Fortune* magazine for six consecutive years, from 1996 to 2001 and was on the *Fortune's* '100 Best Companies to Work for in America' list in 2000, with credible CSR programmes. However, once its greed-fuelled 'financial irregularities' were exposed, Enron's stock price dropped from $90 to cents in the space of 10 months in 2001 and by December, Enron had filed for bankruptcy.

Even as the company was collapsing, greed persisted, with executives taking bonuses of $55 million in the company's last year, while the average staff severance payment was $45,000. Employees lost $1.2 billion in pensions, retirees lost $2 billion, but executives cashed in $116 million in stocks. The dissolution of the complicit Andersen accounting firm resulted in the loss of 85,000 jobs around the world. After a six year class action lawsuit by 1.5 million Enron shareholders, a settlement of $6.79 per share was reached. This was paid from a $7.2 billion compensation fund following class action lawsuits against the banks that did business with Enron, which shareholders allege were aiding and abetting fraud (Visser and CPSL 2009).

Despite these calamitous impacts, the Lehman Brothers story shows that we learned very little from Enron's demise. In June 2006, one of the investment bank's managing directors warned Lehman's analysts that the U.S. real estate market was 'pumped up like an athlete on steroids, rippling with a set of muscles that did not naturally belong there'. Furthermore, that it was based on 'money that was not real money, home prices that were not real prices, and mortgages that were not grounded in any definition of reality' (McDonald 2009).

Despite this insight, Lehman's continued to chase what insider Larry McDonald (2009) called 'one of the greatest consumer borrowing bonanzas since the 1920s', leading to 'America living in a false economy, because all this free money was in defiance of the natural laws of the universe.' And to reward themselves handsomely in the process: Wall Street bonuses were 250 % bigger than the average salary for all nonfinancial jobs in the city, and since 2003, thanks to derivatives, their total compensation had increased by nearly 50 %.

When the 158-year old company filed for Chapter 11 bankruptcy on 15 September 2008, owing $660 billion, it took a good portion of Wall Street, Main Street and the global economy down with it. The financial cost of cleaning up after the global financial crisis—which ultimately gets translated into a tax burden on the public—was estimated by the IMF in August 2009 at $11 trillion. A more recent study by Better Markets (2012) suggests that the U.S. economy alone has lost $12.8 trillion since Lehman Brothers' collapse. And an ILO and OCED (2012) report for the G20 countries found that a 21 million jobs gap has accumulated across the G20 since the onset of the crisis in 2008.

The point to note here is that even Lehman Brothers was savvy to the CSR trend. They issued annual CSR reports and declared to their shareholders in 2007 that: 'Strong corporate citizenship is a key element of our culture. We actively leverage our intellectual capital, network of global relationships, and financial strength to help address today's critical social issues'. They even had an expert in socially responsible business practices join the firm as global head of Sustainability and president of the Council on Climate Change. And bizarrely, in 2008, the firm 'posthumously' received a CSR award for a 10-year mentoring project at a local secondary school in the East End of London (Visser 2011).

Hence, we can see that the age of greed is characterised by defensive CSR in which all corporate sustainability and responsibility practices—which are typically limited—are undertaken only if and when it can be shown that shareholder value will be protected as a result. Hence, employee volunteer programmes (which show evidence of improved staff motivation, commitment and productivity) are not uncommon, nor are defensive expenditures (for example in pollution controls), which are justified in terms of fending off regulation or avoiding fines and penalties.

As with cancer, however, the enabling environment is as important as the greedy cell itself. After all, as I argued in my book *Beyond Reasonable Greed* (Visser and Sunter 2002), a certain measure of selfishness is natural, but it needs to be moderated by norms, rules and cultural taboos that keep its destructive tendencies in check. Greed is not the preserve of a few rogue traders, or money-hungry banks. We were all caught up in its web. Our global financial implosion was (and is) a multi-level phenomenon, incorporating executive greed, banking greed, financial market greed, corporate greed and, ultimately, greed embedded in the capitalist system.

2.2 Charitable CSR

Like greed, charity is probably as old as humanity itself. We find admonitions to generosity, especially by the wealthy, in all the world's major religions, from the *zakat* (wealth tax) in Islam and reciprocity in Confucianism to 'the Golden Rule' in Christianity and the 'wheels of the chariot' philosophy of wealth in Hinduism ('riches revolve from one man to another'). Similarly, in all cultures there are values of solidarity and sharing, like *asistencialismo* (giving for poverty alleviation) in Latin America and *ubuntu* ('I am a person through other people') in southern Africa (Visser and Tolhurst 2010).

Despite these ancient and diverse roots, corporate philanthropy owes much of its modern character to the practices of American's nineteenth century tycoons, such as steel and railroad magnates Cornelius Vanderbilt and Andrew Carnegie. Vanderbilt famously gave away $1 million, the largest charitable gift in American history to that date, to endow what would become Vanderbilt University, named in his honour. The equally rich and generous Carnegie (2007) recorded his philosophy on business in *The Gospel of Wealth*, which elaborated on his three-part dictum: (1) To spend the first third of one's life getting all the education one can; (2) to spend the next third making all the money one can; and (3) to spend the last third giving it all away to worthwhile causes.

Besides these—and at least as iconic as founders of the philanthropy movement—was John D Rockefeller, who made his fortune from oil. He is said to have given away $540 million over his lifetime and died in 1937, aged 98, with a residual estate worth 'only' $26 million. More important even than his individual contribution, he instilled the philanthropic tradition in his family, with his son, 'Junior', giving away over $537 million over his lifetime, and one of his grandsons, David Rockefeller, donating about $900 million to date. A Rockefeller Archive Center study in 2004 documents an incomplete list of 72 major institutions that the family has created and/or endowed up to the present day (Visser 2011).

Continuing the tradition set by Vanderbilt, Carnegie and the Rockefellers are the modern super-rich, notably Bill Gates and Warren Buffet. Microsoft mogul Gates, ranked by *Forbes* magazine as the richest person in the world between 1995 and 2007 and with net worth of around $61 billion in 2012, stunned fans and critics alike when he set up the Bill and Melinda Gates Foundation in 2000 and rapidly grew its assets to a staggering $30 billion by 2007. He was joined by investment tycoon Buffett, who doubled the Gates Foundation assets by gifting over 80 % of his personal wealth. In the spirit of Rockefeller and Carnegie, Gates plans to give away 95 % of his wealth in his lifetime and Buffett plans to leave his children just 'enough to do anything, but not enough to do nothing.'

Other individual philanthropists that have captured the public imagination include CNN founder Ted Turner, Hungarian born Wall Street icon George Soros and British entrepreneur and founder of Virgin, Richard Branson. Less known by the Western public perhaps, but no less successful or generous, are Sheikh

Mohammed bin Rashid al-Maktoum, the ruler of Dubai, and Li Ka-Shing, the Asian tycoon who, among many other business ventures, is the world's largest operator of container terminals.

All of these individuals—and we could name many more—are prototypical philanthropists in the Rockefeller tradition: their charitable activities are funded out of their personal wealth, usually through a foundation bearing their name; their donations are highly public acts, communicated as a legacy statement; and the emphasis is on post-wealth generosity, rather than the ethics (or otherwise) of how they made their money in the first place.

A natural consequence of the individual philanthropy movement was the emergence in the late 1800s in the West of institutional philanthropy, whereby charitable donations are funded directly from business profits, rather than from business leaders' personal wealth. After the World War II, with the increasing proliferation of charities and the professionalisation of corporate philanthropy, it became increasingly common for companies to institutionalise their giving by setting up a corporate foundation, sometimes also called a Chairman's Fund.

In 2012, the top 50 corporate foundations in the U.S. each had assets of over $15 million, with the largest topping $392 million (Foundation Center 2012). Together, their contributions totaled $2.3 billion, a mere fraction of the Bill and Melinda Gates Foundation. Taking a wider sample, the Committee Encouraging Corporate Philanthropy (CECP 2012) which pools data from 214 U.S. companies, including 62 of the top 100 companies in the Fortune 500, reported that corporate giving amounted to $20 billion in 2011. According to the Giving USA (2012), corporate giving accounted for just 5 % of the total giving in 2011 in the United States.

The Rockefeller and Gates stories are good illustrations for the age of philanthropy, because their views on charity embody much of the philanthropic attitudes that still prevail today in business. At the heart of the age—and its chief agent, charitable CSR—is the notion of 'giving back to society'. Rather interestingly, this presupposes that you have taken something away in the first place. Hence, charitable CSR embodies the principle of sharing the fruits of success, irrespective of the path taken to achieve that success. It is the idea of post-wealth generosity, of making lots of money first and then dedicating oneself to the task of how best to distribute it, by way of leaving a legacy.

Although this attitude prevails, Porter and Kramer (2002) have attempted to finesse the concept, arguing that philanthropy can also link to competitiveness. 'Increasingly, philanthropy is used as a form of public relations or advertising, promoting a company's image through high-profile sponsorships', they concede. 'But there is a more truly strategic way to think about philanthropy. Corporations can use their charitable efforts to improve their competitive context—the quality of the business environment in the locations where they operate. Using philanthropy to enhance competitive context aligns social and economic goals and improves a company's long-term business prospects. Addressing context enables a company not only to give money but also leverage its capabilities and relationships in support of charitable causes'.

2.3 Promotional CSR

Research consistently points to a strong marketing driver for CSR (Kotler 2011). According to the CEO survey by Accenture and UN Global Compact (2010), 72 % of CEOs cite 'brand, trust and reputation' as the main factor that has driven them to take action on sustainability issues. Other studies have produced similar results, ranking 'improved company or brand image' as the greatest organisational benefit to addressing sustainability (MIT Sloan Management Review and BCG 2009, 2011).

While these brand and reputational benefits help to build the business case for CSR, they can also lead companies to get stuck in what I call promotional CSR. By this I mean using marketing spin to create an image of responsibility, while failing to change the underlying negative impacts. The tobacco industry is a past master at this. For decades, as research on the negative health impacts of smoking has piled up,[1] the industry sponsored a campaign of disinformation and deception.

This reached its zenith when, in 1994, the CEOs of seven of America's largest tobacco companies testified before the House Subcommittee on Health and the Environment of Congress, all denying that cigarettes are addictive. They lied under oath. Two years later, an investigative article in *Vanity Fair* (Brenner 1996) entitled 'The Man Who Knew Too Much' told the true story of Jeffrey Wigand, a research chemist working for a tobacco industry, who planned to go on the 60 Minutes TV show to expose the lies and deception of the industry, including of the CEOs that he labelled 'The Seven Dwarves'. The story was later turned into the 1996 movie, *The Insider*, starring Russell Crowe as Wigand.

As a result of these and other anti-tobacco campaigns, led by the WHO, tobacco companies have been scrambling to regain their lost credibility and to use CSR to present a more responsible face, seemingly with some success. For example, British American Tobacco (BAT) have engaged in extensive stakeholder consultation exercises and, since 2001, their businesses in more than 40 markets have produced Social Reports, many of which have won awards from organisations as diverse as the United Nations Environment Programme, PwC and the Association of Certified Chartered Accountants. BAT has also been ranked in the Dow Jones Sustainability Index, the FTSE Ethical Bonus Index and Business in the Community (BITC) Corporate Responsibility Index.

Another sector that has been accused of a yawning gap between their PR-massaged image and actual practice is the fossil fuel industry, especially on environmental issues. The classic case today is BP. Despite some progressive action on environmental issues under CEO John Browne in the 1990s, they made

[1] According to the World Health Organisation (WHO), 'no other consumer product is as dangerous, or kills as many people. Tobacco kills more than AIDS, legal drugs, illegal drugs, road accidents, murder and suicide combined.' Of everyone alive today, 500 million will eventually be killed by smoking, and while 0.1 billion people died from tobacco use in the twentieth century, ten times as many will die from tobacco use in the twenty-first century.

the mistake of rebranding in 2000 as ‘Beyond Petroleum’. The company reportedly spent $7 million in researching the new Helios brand and $25 million on a campaign to support the brand change. Greenpeace was not impressed, concluding at the time that ‘this is a triumph of style over substance. BP spent more on their logo this year than they did on renewable energy last year’ (Visser 2011).

Antonia Juhasz (2008), author of *The Tyranny of Oil*, was similarly sceptical, claiming that at its peak, BP was spending 4 % of its total capital and exploratory budget on renewable energy and that this has since declined. In fact, BP’s environmental track record has been dire since they changed their logo, most visibly with the Gulf of Mexico spill in 2010 for which they have set aside $38 billion to cover ensuing liabilities. Besides this, BP has made significant investments in the carbon-intensive Alberta tar sands and continues to attract criticism for their environmental legacy in Nigeria.

This kind of ‘marketing versus reality’ gap on environmental performance is sometimes called ‘greenwash’. The word was coined by environmentalist David Bellamy in the 1980s and plays off of the concept of ‘whitewashing’—literally painting over the cracks to cover up inherent faults. In 1999, the Oxford English Dictionary added the term, defining it as: ‘Disinformation disseminated by an organisation, so as to present an environmentally responsible public image; a public image of environmental responsibility promulgated by or for an organisation, but perceived as being unfounded or intentionally misleading.’

Of course, it is not only BP that is guilty of greenwashing. Another illustrative example from the sector was an advert run by Shell showing a factory with flowers coming out of the smoke-stacks and claiming: ‘We use our waste CO_2 to grow flowers’. There was a grain of truth in the claim, as in the Netherlands the company did capture CO_2 and use it in floral hothouses. However, since Shell only used 0.325 % of its CO_2 output in this way, the Advertising Standards Authority banned the advert, following complaints (Visser 2011).

As a result of this kind of greenwash, the UK’s Committee of Advertising Practice (CAP) Code, enforced by the Advertising Standards Authority, created a clause for environmental claims in 1995. Since 1998, it has also published a non-binding ‘Green Claims Code’, advising advertisers on how best to make good claims. The European Union has also included a clause on greenwashing in their CSR Policy, stating that it will: ‘Address the issue of misleading marketing related to the environmental impacts of products (so-called “green-washing”) in the context of the report on the application of the Unfair Commercial Practices Directive 18 foreseen for 2012, and consider the need for possible specific measures on this issue.’

Of course, this kind of Promotional CSR does not only apply to environmental issues. After the launch of the UN Global Compact, companies started to be accused of ‘bluewash’—a reference to the blue of the UN logo and business using association with the United Nations to appear more responsible than they really are. Likewise, although I haven’t heard the term, I can imagine the ‘redwash’ brush being applied to companies claiming social, community or labour responsibility that masks their real negative impacts on society. The point is not to deny the reputational benefits of CSR, but rather to close the gap between PR claims and actual performance.

2.4 Strategic CSR

Strategic CSR, emerging from the age of management, means relating CSR activities to the company's core business, often through adherence to CSR codes and implementation of social and environmental management systems, which typically involve cycles of CSR policy development, goal and target setting, programme implementation, auditing and reporting. Strategic CSR is the result of a long historical journey, going all the way back to the industrial welfare movement of Victorian times—where, as George Cadbury put it 'the first thought is of the welfare of the work people employed.'

In fact Cadbury's are a prototypical company of the age of management and a showcase of how the practice of strategic CSR emerged over time (Visser 2011). To begin with, the Cadbury brothers are credited with introducing many of England's most progressive workplace practices. One of their most significant actions was to move their factory in 1879 from the grimy city of Birmingham to Bournville in the English countryside—to create a 'factory in a garden'. Working conditions were also progressive. Cadbury's was the first company in England to introduce the five-and-a-half day working week. They were also pioneers in providing medical and dental facilities, offering a pension scheme and shutting the factory on bank holidays (Visser 2013).

Not surprisingly, given their progressive track record on workplace issues, Cadbury's also played a key role in addressing issues of fair trade and supply chain ethics. As far back as 1905, the Cadbury brothers stopped buying cocoa from São Tomé because of poor labour conditions. As a result, they helped found the cocoa industry in Ghana. A 100 years later, Cadbury's launched its first Fairtrade labelled chocolate, which was the culmination of a whole raft of responsible supply chain management initiatives, including the Cadbury Cocoa Partnership (addressing child labour in Ghana and Cote d'Ivoire), the Roundtable on Sustainable Palm Oil (RSPO), the International Cocoa Initiative (ICI) and the Better Sugar Cane Initiative (BSCI).

Prior to their takeover by Kraft, Cadbury's boasted nineteen corporate policies covering various aspects of responsible business practice, from environment, health and safety to marketing, ethics and stakeholder engagement. Beyond these internal commitments, Cadbury's also made very public commitments to responsibility, for example as a signatory to the Courtauld Agreement, Waste and Resources Action Programme (WRAP), the UN Millennium Development Goals (via the Business Call to Action) and the UN Global Compact.

Hence, Cadbury's has—like many other multinationals—caught the wave of Strategic CSR, carried along by the plethora of CSR codes and standards that have emerged since the 1990s, ranging from ISO 14001, OHSAS 18001 and SA 8000, to the Forest Steward Council (FSC), Fairtrade and Ethical Trading Initiative certification schemes, to mention just a few. When we look at the last 10 years, we have seen codes proliferate in virtually every area of sustainability and responsibility and in all major industry sectors. So much so that in *The A to Z of Corporate Social Responsibility*, we included over 100 such codes, guidelines and standards—and that was just a selection of what is out there (Visser et al. 2007).

Beyond the adoption of CSR codes and standards, however, strategic CSR is about focusing CSR initiatives on addressing the material impacts of the company. According to Porter and Kramer (2006), 'Strategic CSR moves beyond good corporate citizenship and mitigating harmful value chain impacts to mount a small number of initiatives whose social and business benefits are large and distinctive' (10). Hence, they conclude, 'The essential test that should guide CSR is not whether a cause is worthy but whether it presents an opportunity to create shared value—that is, a meaningful benefit for society that is also valuable to the business' (8).

Coca-Cola provides an illustrative case of the shift to strategic CSR (Visser 2013). In 2002, residents of Plachimada, a village in India's southern state of Kerala, accused the company's bottling plant there of depleting and polluting groundwater. Two years later, the local government forced Coca-Cola to shut down the plant. In 2006, their situation got worse when a New Delhi research group found high levels of pesticides in Coca-Cola and PepsiCo's locally produced soft drinks, resulting in several Indian states banning their products. Coca-Cola denied any wrongdoing, claiming that borehole water-fed farming was mainly responsible for lowering the water table and that the pollution claims were unsubstantiated. However, the public perceptions battle had already been lost.

Coca-Cola realised that it needs to be seen as part of the solution, not part of the problem. As a result, it has put resources into water at an unprecedented scale. In 2007, the company announced it would spend \$20 million over 5 years to help the WWF preserve seven of the world's major rivers. It also set up the \$10 million Coca-Cola India Foundation, which began installing over 4,000 rainwater harvesting programmes and providing clean drinking water to 1,000 schools across the country.

More significantly, in June of the same year, CEO Neville Isdell flew to Beijing and pledged that his company would become 'water neutral', saying, 'Water is the main ingredient in nearly every beverage that we make. Without access to safe water supply, our business simply cannot exist.' They still have a long way to go—despite improving water efficiency for eight consecutive years, figures show that Coca-Cola replenished, or offset, only 23 % of the water used in production in 2010. Nevertheless, they are driving improvement, as is typical for strategic CSR, by having clear objectives and targets linked to water, and tying these to a sustainability-related performance bonus (Crognale 2012).

This process of continuous improvement through a management systems approach is one of the strengths of strategic CSR—and also one of its weaknesses, as we will explore in the section on the failure of CSR 1.0. A final distinction to make is that strategic CSR results in various social or environmental issues being tackled which are aligned with the company's core business, but it seldom results in the company changing its strategy or core business.

2.5 Transformative CSR

Transformative CSR, or CSR 2.0, in an age of responsibility focuses its activities on identifying and tackling the root causes of our present unsustainability and irresponsibility, typically through innovating business models, revolutionising their processes, products and services and lobbying for progressive national and international policies. Hence, while strategic CSR is focused at the micro level—supporting social or environmental issues that happen to align with its strategy (but without necessarily changing that strategy)—transformative CSR focuses on understanding the interconnections of the macro level system—society and ecosystems—and changing its strategy to optimise the outcomes for this larger human and ecological system.

When Ray Anderson, founder of Interface FLOR, formulated a new vision for his carpet tile company in 1994, he unwittingly became a pioneer of transformative CSR. It began with the realisation that carpet manufacturing 'is a pretty abusive industry'. The process uses lots of petroleum and petroleum derivatives, both as components of synthetic carpet and to power its production. Dyeing carpet is also water- and energy-intensive. And when people are finished with the carpet, it goes into landfills where it lasts probably 20,000 years.

Anderson concluded that his company—and business more generally—is part of the problem, not the solution. To paraphrase Paul Hawken (1994), he realised that 'there is not an industrial company on earth, not an institution of any kind, not mine, not yours, not anyone's, that is sustainable.' And so Anderson crystallised his vision: Interface would become the world's first truly sustainable company. In fact, not only sustainable, but restorative. They would put back more than they take, and actively do good, not just avoid doing harm.

This is what makes Ray Anderson and Interface FLOR different from, say, BP or Cadbury's. It is the depth of their admission and the scale of their ambition. Anderson's (2009) last book was called *Confessions of a Radical Industrialist*, in which he conceded not only that our modern economic system is broken, but that he and his company were part of the problem. He was able to see himself as (to use his own words) 'a plunderer'—not through malicious intent, or even greed, but by failing to question the true impacts of business on society and the environment.

As Alcoholics Anonymous will tell you, admission is the first step to recovery. Unfortunately, most companies stuck in the ages of greed, philanthropy, marketing and management are all still in denial, thinking that either there is *no* problem, or it's not *their* problem, or that it's a problem to *benefit* from, or that it's only a *minor* problem.

The age of responsibility is not just about admission though; it's also about ambition. As far as I can tell, Interface FLOR was the first major company to set the BHAG (big hairy audacious goal) of zero negative impact, as well as going beyond 'no harm' to also become a restorative business—to genuinely make things better and leave this world with a net-positive balance. Today, Interface FLOR calls this 'mission zero', or 'mission sustainability', which Anderson always likened to summiting 'a mountain higher than everest'—difficult, yes, but with a careful and attentive plan, not impossible.

Importantly, their ambitious goal of zero impact is fuelled by performance measures; what Interface FLOR calls EcoMetrics. The numbers show that Interface FLOR is on track to meet their 2020 zero impact goal. Since 1996, per unit of production, waste to landfill is down 88 %, water use is down 84 %, energy use is down by 47 % and non-renewable energy is down by 64 %. Furthermore, the company has achieved an absolute reduction on greenhouse gas (GHG) emissions of 32 %, while 31 % of global energy used is from renewable sources and 44 % of total raw materials are recycled or bio-based materials.

Besides admission and ambition, Interface FLOR is a good example company for transformative CSR because they have used innovation to tackle the root causes of our unsustainable economic system, namely our take-make-waste model of industrial production. For example, in 1995, Interface FLOR launched an evergreen lease service, where the company produces, installs, cleans, maintains and replaces the carpet for customers, thereby ensuring effective take-back and recycling.

Other product innovations have included the Cool Carpet (offsetting greenhouse gas emissions) and carpets using fly ash waste and polylactic acid (PLA) fibres, derived from non-food grade corn. Perhaps most famously, in 2006, Interface also invented the world's first totally glue-free, free-lay carpet tile (TacTile®), inspired by gecko-foot 'technology'. The result is less mess, less waste and greater savings, not to mention an environmental footprint that is over 90 % lower than carpets using traditional glue adhesives.

Anderson was not the first radical business leader, nor perhaps even the most radical. Anita Roddick, founder of The Body Shop International, had a missionary zeal that few will ever rival. Famous for her business-led activism, which began as an alliance with WWF in 1986 to save the whale, she went on to tackle issues as far ranging as animal rights, women's self-esteem, human rights, fair trade and indigenous people's rights. In her autobiography, *Business As Unusual*, Roddick (2001) distilled her philosophy as follows: 'Business is a renaissance concept, where the human spirit comes into play. It does not have to be drudgery; it does not have to be the science of making money. It can be something that people genuinely feel good about, but only if it remains a human enterprise.'

There are many other examples of companies pioneering the transformative CSR approach, from small enterprises like Seventh Generation and Green Mountain Coffee to multinational behemoths like Nike and Unilever (Visser 2013). Some of these, I will explore in more detail in the sections to follow. However, none of them are perfect, and many perform poorly in some areas, even while excelling in others. What qualifies them as CSR 2.0 leaders is that they are using CSR as a catalyst for systemic change (Eisenbeiss 2012). Not only are they changing their own corporate strategies, they are also working to shift the practices of their industry sector, as well as lobbying for policy reform and cultural values that are inherently more responsible and sustainable.

We will now look at how the CSR 2.0 concept emerged and why it is such a powerful and relevant metaphor for transforming CSR theory and practice.

References

Accenture and UN Global Compact (2010) A new era in sustainability [CEO survey]
Anderson RC (2009) Confessions of a radical industrialist. Random House Business Books, London
Better Markets (2012) The cost of the wall street-caused financial collapse and ongoing economic crisis is more than $12.8 Trillion, 15 Sept
Brenner M (1996) The man who knew too much. Vanity fair, May. Harvard Business School Press, Cambridge
Carnegie D (2007) The gospel of wealth. Standard Publications, New York (Originally published in 1889)
CECP (2012) 2012 Giving in numbers: trends in corporate giving. Report
Crognale G (2012) How coca-cola guides a thousand plants to improve sustainability. Sustainable plant, 12 Oct
Eisenbeiss SA (2012) Re-thinking ethical leadership: an interdisciplinary integrative approach. Leadersh Quart 23(5):791–808
Foundation Center (2012) 50 largest corporate foundations by total giving. http://foundationcenter.org/findfunders/topfunders/top50giving.html. Accessed 18 Dec 2012
Giving USA (2012) 2012 Giving USA: the annual report on philanthropy for the year 2011
Hawken P (1994) The ecology of commerce: a declaration of sustainability. HarperBusiness
ILO and OCED (2012) Short-term labour market outlook and key challenges in G20 countries. Statistical update prepared by the ILO and the OECD for the meeting of G20 labour and employment ministers Guadalajara, 17–18 May 2012
Juhasz A (2008) The tyranny of oil: the world's most powerful industry – and what we must do to stop it. New York: William Morrow
Kotler P (2011) Reinventing marketing to manage the environmental imperative. J Mark 75(4):132–135
McDonald L (2009) A colossal failure of common sense: The Incredible Inside story of the collapse of Lehman brothers. Crown Business, New York
MIT Sloan Management Review and BCG (2009) The business of sustainability. Special Report
MIT Sloan Management Review and BCG (2011) Sustainability: The 'Embracers' Seize Advantage. Special Report
Porter ME, Kramer MR (2002) The competitive advantage of corporate philanthropy. Harvard Bus Rev 80:57–68
Porter ME, Kramer MR (2006) Strategy and society: the link between competitive advantage and corporate social responsibility. Harvard Bus Rev 84(12):56–68
Roddick A (2001) Business as unusual. Thorstens, London
Visser W (2010) CSR 2.0: from the age of greed to the age of responsibility. In: Sun W et al (eds) Reframing corporate social responsibility: lessons from the global financial crisis. Bingley, Emerald

Visser W (2011) The age of responsibility: CSR 2.0 and the new DNA of business. John Wiley & Sons London
Visser W (2012) The quest for sustainable business: an epic journey in search of corporate responsibility. Greenleaf, Sheffield
Visser W (2013) Corporate sustainability and responsibility: an introductory text on CSR theory and practice – past, present and future. Kaleidoscope Futures, London
Visser W, Sunter C (2002) Beyond reasonable greed: why sustainable business is a much better idea!. Tafelberg Human & Rousseau, Cape Town
Visser W, Tolhurst N (2010) The world guide to CSR. Greenleaf, Sheffield
Visser W and CPSL (Cambridge Programme for Sustainability Leadership) (2009) Landmarks for sustainability: events and initiatives that have changed our world. Greenleaf Publishing, Sheffield
Visser W, Matten D, Pohl M, Tolhurst N (eds) (2007) The A to Z of corporate social responsibility: a complete reference guide to concepts, codes and organisations. Wiley, London
Zadek S (2004) The path to corporate responsibility. Harvard Bus Rev 82(12):125–132

Chapter 3
The Failure of CSR 1.0

Abstract Trends are presented for the most serious global social, environmental and ethical challenges. The failure of CSR 1.0 approaches (the first four stages) to have any significant impact on these trends is explained in terms of three factors, namely the incremental, peripheral and uneconomic nature of CSR 1.0 approaches.

Keywords Ecosystems · Environment · Climate change · Poverty · Inequality · Corruption · Total quality management · Business case · Globalization · Transparency

Before I explain why CSR 1.0 has failed, let me share some data to back up my claim that many of our global problems are getting worse, not better, despite more CSR than ever before.

3.1 Ecosystems Decline

According to the Global Footprint Network, humanity's ecological footprint, driven by the spread of capitalism and Western lifestyles globally, has more than tripled since 1961. Since the late 1980s, we have been in 'overshoot'—meaning that the world's ecological footprint has exceeded the Earth's biocapacity. An ecological footprint analysis shows that global biocapacity—the area available to produce our resources and capture our emissions—is 1.8 global hectares (ha) per person, while the footprint per person is already 2.7 global ha (2007 figures).

The Global Footprint Network calculates that our demand for renewable ecological resources and the services they provide is now equivalent to more than 1.5 Earths. The data shows us on track to require the resources of two planets well before mid-century. In 2012, if everyone on the planet lived the high-consumption lifestyles of the USA, we would need 4.16 Earths. Comparative statistics are

W. Visser, *CSR 2.0*, SpringerBriefs in Organisational Studies,
DOI: 10.1007/978-3-642-40874-8_3, © The Author(s) 2014

available for around 150 countries, including the United Arab Emirates (5 Earths), UK (2.6), Germany (2.57), Russia (2.48), Japan (2.35), Brazil (1.65), South Africa (1.46), Russia, China (1.2) and India (0.49).

A second environmental indicator is the Living Planet Index, which reflects changes in the state of the planet's biodiversity, using trends in the size of 9,014 populations of 2,688 mammal, bird, reptile, amphibian and fish species from different biomes and regions. The Living Planet Index continues to show a 28 % global decline in biodiversity health since 1970, with species in the tropical region having declined by more than 60 % from 1970 to 2008. These dramatic losses in our natural wealth are being driven by deforestation and land conversion in the tropics and the impact of dams, diversions and climate change on freshwater species (Ollivier 2012). Pollution, over-fishing and destructive fishing in marine and coastal environments are also taking a considerable toll.

Another indicator of the state of the planet is the UN Millennium Ecosystem Assessment, issued in 2005, which reaches similar conclusions: 60 % of world ecosystem services have been degraded; of 24 evaluated ecosystems, 15 are being damaged; water withdrawals have doubled over the past 40 years; over a quarter of all fish stocks are overharvested; since 1980, about 35 % of mangroves have been lost; about 20 % of corals have been lost in just 20 years and 20 % more have been degraded; and species extinction rates are now 100–1,000 times above the background ('natural') rate. So, by all accounts, capitalism is failing spectacularly to control the environmental impacts of the very economic activities that it is so successful at stimulating (Whiteman et al. 2013; Böhm et al. 2012).

What many people fail to appreciate is how uneconomic this environmental destruction really is. For example, a 2010 study conducted for the UN by Trucost found the estimated combined damage of the world's 3,000 biggest companies was worth \$2.2 trillion in 2008—a figure bigger than the national economies of all but seven countries in the world that year, and equal to one-third of the average profits of those companies. In 2010, The Economics of Ecosystems and Biodiversity (TEEB) study led by Pavan Sukhdev also found that degradation of the Earth's ecosystems and biodiversity due to deforestation alone costs us natural capital worth somewhere between \$1.9 and \$4.5 trillion every year.

3.2 Climate Change

Our environmental impacts and associated economic costs are no more dramatically evident than on the issue of climate change. The 4th Assessment Report of the Intergovernmental Panel on Climate Change (IPCC 2007) concluded that global atmospheric concentrations of greenhouse gases (GHGs) have increased markedly since 1750 as a result of human activities. Ice-core records spanning thousands of years show that concentrations today far exceed recent historical levels, with carbon dioxide (CO_2, the most important GHG) growing from 280 parts-per-million (ppm) in pre-industrial times, to 379 ppm by 2005. This exceeds the natural range over the last 650,000 years.

According the Global Footprint Network, the world's carbon footprint is 54 % of humanity's overall ecological footprint and represents its most rapidly-growing component. Humanity's carbon footprint has increased 11-fold since 1961. Moreover, the rate of increase in CO_2 concentration has been faster in the last decade than at any point since measurement began. The spike in carbon emissions is mainly due to fossil fuel use, although changes in land-use are also a big factor. Other GHG concentrations (e.g. methane) have also been increasing.

Despite occasional media-grabbing stunts by climate sceptics, there is overwhelming scientific consensus that the climate system is warming and that human activity is the main cause. Trends since 1900 indicate significantly increased precipitation in areas such as northern Europe and drying in areas like the Mediterranean. Longer, more intense droughts have been seen since the 1970s and there have been widespread changes in extreme temperatures over the last 50 years.

The most recent 100-year linear trend shows a 0.74 °C increase in temperature in the century to 2005. Overall, the sea level is estimated to have risen by 0.17 m during the twentieth century. A warming of 0.2 °C per decade over the next 20 years is predicted and there is a greater than 90 % chance that climate changes during the twenty-first century will exceed the previous century. The current best estimate for the average temperature rise is 1.8–4.0 °C by 2100, with a possible range of 1.1–6.4 °C. It is widely held that any rise above 2 °C risks 'runaway', 'dangerous' or 'catastrophic' climate change, where self-reinforcing feedback loops in the climate system cause rising temperatures to spiral out of control, at which point human mitigation becomes ineffectual.

Specific predictions are also possible. For example, snow cover is expected to decrease and permafrost regions (which store vast amounts of methane) will likely see increases in thaw depth. Temperature extremes, heat waves and heavy precipitation events will continue to become more frequent and tropical cyclones become more intense. The higher latitudes will probably see more precipitation and most subtropical regions less. In addition, it is likely that human activities have increased the risk of heat waves.

The Stern Review on *The Economics of Climate Change* concludes that climate change is 'the greatest market failure the world has ever seen' and estimates that the cost of action to reduce GHGs and avoid the worst impacts of climate change can be limited to about 1 % of global GDP per year if action is immediate and decisive. By contrast, failure to act swiftly will damage economic growth. Specifically, inaction would result in a persistent annual loss of 5 % of global GDP. If a wider range of impacts and risks is considered, this could be as high as 20 % of GDP, or more (Stern 2006).

It is important to emphasise that climate change is not just an environmental issue. The UNDP (2009) estimates significant impacts of global warming on the world's 2.6 billion people surviving on less than $2 a day, including up to 600 million more people facing malnutrition due to the breakdown of agricultural systems resulting from increased exposure to drought, rising temperatures, and more erratic rainfall. Potential productivity losses of 26 % are expected by 2060 in semi-arid areas of sub-Saharan Africa, home to some of the highest concentrations of poverty in the world.

Furthermore, an additional 1.8 billion people are expected to experience water stress by 2080, with large areas of South Asia and northern China facing a grave ecological crisis as a result of glacial retreat and changed rainfall patterns. In addition, up to 332 million people in coastal and low-lying areas may be displaced by flooding and tropical storm activity, including more than 70 million Bangladeshis, 22 million Vietnamese, and 6 million Egyptians. Finally, related health effects suggest that as many as 400 million people are likely to face the risk of malaria as a result of climate change.

3.3 Poverty

The social impacts of our globalisation activities are more ambiguous. On the one hand, critics like Naomi Klein (2000, 2007), author of *No Logo* and *The Shock Doctrine*, argues that 'Gucci capitalism' results in labour exploitation and a 'race to the bottom'. In other words, capital flows to wherever the social or environmental standards are lowest. The critics go further to suggest that capitalism is designed to create the instability that we have seen in the markets, and those that suffer the most from this volatility are always the most vulnerable, namely the poor of the world (Stiglitz 2010).

On the other hand, there has been undoubted progress in reducing global poverty. The UNDP (2012) Millennium Development Goals (MDGs) report shows that, for the first time since records on poverty began, the number of people living in extreme poverty has fallen in every developing region, including sub-Saharan Africa. The proportion of people living on less than $1.25 per day fell in 2010 to less than half the 1990 rate and during the same period over two billion people gained access to improved drinking water sources. The share of slum dwellers in urban areas declined from 39 % in 2000 to 33 % in 2012, improving the lives of at least 100 million people.

Besides this, primary school enrolment of girls equalled that of boys, and we have seen accelerating progress in reducing child and maternal mortality. The number of under-five deaths worldwide fell from more than 12 million in 1990 to 7.6 million in 2010. We have also seen success in tackling global diseases. Tuberculosis incidence rates have been falling since 2002, and current projections suggest that the 1990 death rate from the disease will be halved by 2015. At the end of 2010, 6.5 million people were receiving antiretroviral therapy for HIV or AIDS in developing regions, and the estimated incidence of malaria has decreased globally by 17 % since 2000.

Despite this remarkable progress, however, huge challenges remain. Vulnerable employment accounted for an estimated 58 % all employment in developing regions in 2011, affecting women and youth most acutely. Nearly half of the population in developing regions—2.5 billion—still lacks access to improved sanitation facilities, and the number of people living in slums continues to grow to an estimated 863 million. Furthermore, the FAO estimates that 850 million people (15 % of the world's population) are still living in hunger. Progress has also been slow in reducing child under-nutrition, with close to a third of children in Southern Asia deemed underweight in 2010.

3.4 Inequality

A recent review of global inequality by UNICEF (Ortiz and Cummins 2011) reveals a world in which the top 20 % of the population enjoys more than 70 % of total income, contrasted by 2 % for those in the bottom 20 % (2007 figures under PPP-adjusted exchange rates). They estimate that it would take more than 800 years for the bottom billion to achieve 10 % of global income under the current rate of change. Also disturbing is the prevalence of children and youth among the poorest income quintiles, as approximately 50 % are below the $2/day international poverty line.

The picture in the U.S. is even starker. According to the Economic Policy Institute (2012), in 2010, the highest-income 1 % of households received 17 % of all income in the economy, while the wealthiest 1 % held 35 % of all wealth. The bottom 90 % meanwhile received only 55 % of all income and held just 23 % of all wealth. From 1983 to 2010, 38 % of the wealth growth went to the top 1 % and 74 % to the top 5 %. The bottom 60 %, meanwhile, suffered a decline in wealth. Even more dramatically, between 1979 and 2007, incomes for the top 1 % grew 241 %, compared to 11 % for the bottom fifth and 19 % for the middle fifth. In 1962, the wealthiest 1 % had 125 times the wealth of a median household. In 2010, the ratio was 288-to-1. From 2007 to 2010, the average wealth of the top 1 % dropped 16 %, while median wealth plummeted 47 %.

This is not only a U.S. phenomenon. Middle-income countries appear the most unequal. Gini Index trends show that Eastern Europe (including former Soviet Union states) and Asia had the largest increases between 1990 and 2008. Latin America remains the region with the highest level of income inequality, despite improvement since 2000. Similarly for sub-Saharan Africa, which is highly unequal but has reduced its Gini Coefficient since 1990.

Inequality between urban and rural populations in developing countries is especially worrying, as access to water and sanitation illustrates. According to the UNDP (2012), over 90 % of the richest fifth of the urban population of sub-Saharan Africa use improved water sources, and over 60 % have piped water on premises. By contrast, in rural areas, piped-in water is non-existent in the poorest 40 %, and less than half of the population use any form of improved source of water. Similarly, over 90 % of the households in the richest urban quintile benefit from improved sanitation, while access in rural areas falls below 50 % even among the wealthiest households. In the poorest rural quintile, over 60 % of households practice open defecation.

Beyond development indicators, inequality also exists in the distribution of subjective happiness in the world. For example, Gallup's (2010) global wellbeing poll revealed that the percentage of people who are 'thriving' (as opposed to 'struggling' or 'suffering') ranges from a high of 82 % in Denmark to a low of 1 % in Togo. Africa has the lowest perceived wellbeing, with no country in this region showing a thriving indicator higher than 25 %. In fact, of the 41 countries where thriving is 10 % or lower, more than half are in Africa. Elsewhere in the world, however, disparities also exist. Thriving in the Americas is highest in Costa Rica (63 %) and lowest in Haiti (4 %); in Europe, it ranges from Denmark (82 %) to Bulgaria (6 %) and in Asia Pacific, from New Zealand (63 %) to Cambodia (3 %).

Inequality is not only a global or national phenomenon; it is also a corporate issue, as we see when we look at trends in relative pay (Economic Policy Institute, 2012). Between 1979 and 2007 in the U.S., wages for the top 1 % rose almost 10 times as fast as those for the bottom 90 %, by 156 % versus 17 %. These disparities are especially pronounced in the explosion of CEO pay, which in the late 1970s was about 30 times that of a typical worker. Today, it is more than 200 times that of a typical worker. From 1978 to 2011, CEO compensation grew more than 725 %. Private-sector worker compensation grew only 6 %.

The situation is not much better in the UK. According to a survey of FTSE 100 companies by Manifest/MM&K (2012), the median increase in chief executive pay in 2011/12 was 10 % (and in 25 companies, more than 40 %), while employees got mean average rises of 1 %. The average remuneration awarded to chief executives across the FTSE 100 was £4.8 million.

3.5 Corruption

One of the socio-economic cancers that aids and abets the poverty and inequality just described is corruption. According to Transparency International's 2012 Corruption Perceptions Index (CPI)—which is a measure of domestic, public sector corruption—two thirds of the 176 countries included score below fifty on a scale from 0 (perceived to be highly corrupt) to 100 (perceived to have low levels of corruption). Fragile, unstable states that are scarred by war and ongoing conflict linger at the bottom of the index. These include Somalia and Afghanistan (scoring just 8, along with North Korea), Sudan (13) and Myanmar (15). Highest scorers are Denmark, Finland and New Zealand (all scoring 90).

Transparency International's 2011/2012 Global Corruption Barometer found that, of the more than 100,000 people polled in 100 countries, 24 % reported having paid a bribe in the previous 12 months and 58 % believe that corruption has increased in the past 3 years, with only 16 % saying it had decreased. In terms of institutional corruption, 68 % believe political parties are corrupt or extremely corrupt, as compared with 41 % for business and 24 % for NGOs. Only 6 % of those surveyed trusted business most to fight corruption, while in Austria, China,

Denmark, Hong Kong, Luxembourg, Netherlands, Norway, Switzerland and Turkey, business is seen as the most corrupt institution.

The Barometer also found that the poorest families continue to be punished by petty bribe demands. Across the board, low-income respondents were more likely to be met with bribe demands than high-income respondents. Furthermore, only three in ten respondents believed their government's efforts to fight corruption were effective. Despite this institutional malaise, 70 % think that ordinary people can make a difference in the fight against corruption and 75 % would be prepared to report an incident of corruption.

According to another of Transparency International's indexes—the 2011 Bribe Payers Index (BPI)—companies based in the emerging economic giants are perceived to routinely engage in bribery when doing business abroad. For example, Russia ranked last with a score of 6.1 (where 10 represents no corruption), just below China (6.5), Mexico (7.0) and Indonesia (7.1). At the other end of the spectrum, The Netherlands and Switzerland shared first place with a score of 8.8, while third place went to Belgium (8.7). The sectors most likely to pay bribes are Public Works Contracts and Construction (5.3), Utilities (6.1), Real Estate, Property, Legal and Business Services (6.1), Oil and Gas (6.2) and Mining (6.3).

In a survey of 3,016 senior business executives in 30 countries around the world, Transparency International (2011) found that 27 % believed they had lost a competitive bid because a competitor paid a bribe (rising to 50 % in Malaysia, 48 % in Mexico and 47 % in Indonesia). Seventy-nine percent believe companies have an ethical duty to fight corruption and 75 % claim they would report an incident. Forty-three percent have measures in place to support potential whistle blowers, while 68 % claim to prohibit facilitation payments. Fifty-six percent believe their company anti-corruption measures are effective, while only 38 % believe national anti-bribery laws are effective.

Transparency International estimates that bribery, cartels and other corrupt practices undermine competition and contribute to a massive loss of resources for development in all countries, especially the poorest ones. For example, between 1990 and 2005, more than 283 private international cartels were exposed that cost consumers around the world an estimated $300 billion in overcharges.

3.6 The Limitations of CSR 1.0

These bewildering facts and figures leave us with many troubling questions. In this book, I am mainly concerned with those that involve business. For instance, I wrestle with the central question: Are companies more a part of the problem or the solution? Is the net impact of business positive or negative? And given that CSR has increased dramatically over the same 50 years that many of the global problems described above have been getting worse, does that mean that CSR is ineffective?

Even worse, could the CSR boom be an unwitting accomplice to the spate of corporate crimes of recent decades? Are we quietly and unintentionally aiding and abetting our collective demise? After all, Enron, Lehman Brothers and BP all had their fair share of CSR—from codes of conduct and ethics officers to corporate volunteering and community development programmes. And yet, all the CSR programmes in the world did little or nothing to change the internal culture of greed, negligence or narrow pursuit of profits that was nurtured and rewarded over decades by these organisations.

If CSR is used to legitimise businesses or practices that are, in essence, irresponsible (Lange and Washburn 2012), surely CSR is partly to blame for the various corporate 'sins' that go undetected and unpunished? This leads to a very uncomfortable conclusion. At worst, CSR in its most primitive form may be a smokescreen covering up systemically irresponsible behaviour. At best, even the most evolved CSR practices might just be a band-aid applied to a gaping wound that is haemorrhaging the lifeblood of the economy, society and the planet.

The reason that CSR 1.0—in its defensive, charitable, promotional and strategic manifestations—is failing to solve our most pressing global challenges is that it suffers from three fundamental limitations: it tends to be peripheral, incremental and uneconomic, as summarised in Table 3.1 and briefly explained in the sections.

3.7 Peripheral CSR

The first of the problems of CSR 1.0 is the problem of peripheral CSR. Returning to our BP example, this was a company with a long and mostly proud history, contributing highly useful products to society and practicing sophisticated CSR management. Leaving the safety and environmental disasters aside for a moment, BP has made serious commitments to sustainability and responsibility and achieved a great deal in terms of measurable improvements in its safety, health, environmental, labour and human rights performance. And yet for all their flagship leadership in the age of management, we see that CSR has remained on the periphery. BP has not gone 'beyond petroleum'; quite the opposite in fact.

It is the same for almost all companies practicing CSR. At worst—and I see this especially in developing countries that are stuck in philanthropic or promotional CSR mode—CSR sits in a public relations, marketing, corporate affairs or human resources department. It is an 'add-on', explicitly used to improve brand equity or the company's reputation. At best—and more common in developed countries and among subsidiaries of multinationals—we see companies practicing strategic CSR, trying to align CSR activities with their industry impacts, or embedding CSR through management systems (Campbell et al. 2012). Even so, they completely fail to change the strategic direction or core business of the company, or the harmful effects of its processes, products and services.

What Enron, BP, Coca-Cola and virtually every other CSR 1.0 company have in common is not the deliberate intention to mislead (although there are clear

Table 3.1 The Limitations of CSR 1.0

Limitation	Nature of the failing
Peripheral CSR	CSR has remained largely restricted to the largest companies, and mostly confined to PR, or other departments, rather than being integrated across the business
Incremental CSR	CSR has adopted the quality management model, which results in incremental improvements that do not match the scale and urgency of the problems
Uneconomic CSR	CSR does not always make economic sense, as the short-term markets still reward companies that externalise their costs to society

examples of this too), but rather a corporate culture—supported by a system of narrow institutional performance incentives, short-term market pressures and perverse economic measures of progress—that remains essentially in conflict with the objectives of sustainability and responsibility. When a trade-off has to be made between financial profitability and ethical standards, the choice is clear, irrespective of carefully crafted codes of practice on the boardroom wall. If there is a tug-of-war between economic growth and environmental impacts, the winner is clear, despite any number of ISO 14001 certificates. If customer demand for cheap products is at odds with fair labour conditions, consumerism triumphs over the needs of powerless workers in the supply chain from some far-flung land.

Examples are not hard to find and regularly break as corporate scandals in the news (Stolowy 2012). In 2012 alone, HSBC and Standard Chartered were found to be complicit in money laundering (they are to pay more than $2.5 billion in fines as part of record settlements with U.S. authorities). There was also the $1.7 billion accounting cover-up by Olympus in Japan, the Barclays interest-rate-fixing LIBOR scandal (for which it was fined $450 million) and the $24 million 'illicit payment' Walmart bribery scandal (investigations are still on going at the time of writing), not to mention ongoing fallout from the exposure of labour conditions at Foxxcon in China, which implicates Apple, HP and Dell. Activist organisations like Corporate Watch make it their business to profile these and less high profile 'corporate crimes' in the sectors and companies where they manifest.

CSR has remained peripheral in another way. It hardly ever extends beyond the large, high-visibility branded companies in any country. All the CSR indexes and rankings, the CSR codes and standards, the CSR reports and audits are focused on a few thousand companies. By the end of 2012, the Global Reporting Initiative included around 5,000 organisations in its database of sustainability reporters. SA 8000 certification claims to have been applied in 3,000 factories. The UN Global Compact has around 7,000 business signatories. These numbers are peripheral by any measure you care to choose. Even ISO 14001, with 250,000 certifications worldwide by December 2010, pales into insignificance when you consider that the U.S. Chamber of Commerce alone claims to be 'representing the interests of more than 3 million businesses'. If we are honest, CSR is the preserve of a tiny corporate elite, a miniscule business minority.

3.8 Incremental CSR

Closely linked with the peripheral problem—and driven by the age of management—is the problem of incremental CSR. To fully appreciate this issue, we have to go back to business guru Peter Drucker's 1954 book *The Practice of Management*, in which he introduced the concept of 'management by objectives', or MBOs (Drucker 1993). The concept is so endemic now as to seem like common sense, but it was quite a revolutionary concept at the time. The basic idea is to translate corporate strategy into a series of measurable objectives, which can be cascaded down through the organisation. This allows managers to track and incentivise performance, while employees know what is expected of them and can reap the rewards if they meet their targets. Furthermore, if they participate in setting those objectives, they are likely to feel more motivated and empowered.

The MBO approach—together with subsequent tools like the Balanced Scorecard—is right at the heart of the age of management, in the sense that draws attention to voluntary incremental improvements, which distracts attention from the larger problems and deeper impacts of the business. In one of those bizarre ironies of history, the 'system' that would do more to embed the MBO approach than anything else was conceived by one of MBO's great detractors. I am referring to W. Edwards Deming and his total quality management (TQM) approach. Deming credits the inspiration for his theory of management to a 1927 meeting with Walter A. Shewhart of the Bell Telephone Laboratories, the originator of the concepts of statistical control of processes. Years later, during Allied occupation of Japan, Deming was asked by the U.S. military to assist with the 1951 Japanese Census.

This led to an invitation by the Japanese Union of Scientists and Engineers (JUSE) for Deming to teach statistical control and quality management to its members. Japan's CEOs were impressed with Deming's idea that improving quality would reduce expenses, while increasing productivity and market share, and began to test and implement TQM in their factories, notably in their nascent motor industry. Not only did this assist Japan's economic rise in the second half of the twentieth century, but it also spawned the international quality movement.

The TQM approach was later standardized through ISO 9001, first launched in 1987. By the end of 2010, over a million certifications had been issued. The key to total quality management, according to ISO 9001, is continuous improvement, which is predicated on setting objectives and reviewing performance against them. The designers of the standard seem to have overlooked (or ignored) Deming's objection to MBOs. Deming argued that a lack of understanding of systems commonly results in the misapplication of objectives. By contrast, a leader with an understanding of systems was more likely to guide workers to an appropriate solution than the incentive of an objective.

This debate is important for the responsibility debate because the most widely practiced CSR standards, ISO 14001 and more recently ISO 26000 (Helms et al. 2012), are explicitly designed to apply the ISO 9001 approach to management systems, including MBOs, to environmental and stakeholder management. That is not a bad thing in and of itself, and it has resulted in many welcome incremental improvements in the environmental and social performance of companies *processes*. But the Achilles heel of ISO 14001 and all the other voluntary CSR standards that use MBOs is this: companies set their own objectives and make progress at their own pace and discretion. Furthermore, as with the peripheral curse, the MBOs approach has failed to challenge or significantly change companies' largest negative impacts, which are associated with either the nature of their business, the consumption-driven lifestyle they promote, or the impacts of their resource- and energy-intensive products and services.

The net effect is that, despite more CSR than ever before, and despite laudable incremental improvements in CSR performance at the micro level, virtually every macro-level indicator we have of social, environmental or ethical quality—be it the gap between rich and poor, deforestation, biodiversity loss, or corruption—shows that things are still getting worse, not better. The incremental approach to CSR simply does not produce the scale and urgency of response that is required, nor does it get to the root of business's systemic unsustainability and irresponsibility in the shareholder-driven, growth-obsessed capitalist global economy.

3.9 Uneconomic CSR

The third and final limitation of CSR 1.0 is that the much touted 'business case' for CSR is not nearly as obvious, certain or practiced as many assume (Barnett and Salomon 2012). Let's start with the rhetoric. The World Business Council for Sustainable Development (WBCSD), which is the strongest proponent of the business case, suggests that it is predicated on five 'returns': operational efficiency, risk reduction, recruitment and retention of talent, protecting the resource base of raw materials, and creation of new markets, products and services. And it is certainly not hard to find ad-hoc examples of each of these 'win-wins' (Haanaes et al. 2013). But is there always a business case?

To answer this, we must look beyond the rhetoric and turn to academic research. The findings vary. For example, Griffen and Mahon (1997) reviewed 25 years of studies and found that a majority showed a positive link between CSR and financial performance, while Margolis and Walsh (2001) reviewed 80 studies, of which 42 show a positive relationship, 19 demonstrate no relationship and four find a negative one.

Two reports by SustainAbility—*Buried Treasure* and *Developing Value*—also suggest mixed results (SustainAbility and UNEP 2001; SustainAbility, IRC and Ethos Institut 2002). Some relationships between sustainability factors and business success factors are stronger than others, and in many cases, no relationship exists. Laffer et al. (2004), on the other hand, in a review of *Business Ethics* magazine's 100 Best Corporate Citizens found 'no significant positive correlation between CSR and business profitability as determined by standard measures'.

Despite these contradictions, academics continue to pursue the holy grail of instrumental business case relationships in CSR (Tang et al. 2012; Walls et al. 2012; Yongtae et al. 2012). Vogel (2005) believes that 'there is no definitive answer to the question of a financial link. It depends on an individual company's circumstances. Academics searching for a definitive corporate responsibility-financial performance link are barking up the wrong tree.' Carroll and Shabana (2010) agree, saying that 'mediating variables and situational contingencies affect the impact of CSR on firm financial performance. Therefore, the impact of CSR on firm financial performance is not *always* favourable' (102).

Simply put, there are far too many variables to isolate the impact of CSR on financial performance, except through very specific examples like eco-efficiency (Chien-Ming and Delmas 2012). We must also ask are typical measures of CSR a reliable proxy for sustainability and responsibility? After all, if we had correlated Enron's CSR and financial performance prior to its demise, it would have pointed to a strong positive relationship, which makes a nonsense of the whole exercise.

I have a more fundamental problem with the CSR business case rhetoric however. The real question we should be asking is: Does the market consistently reward sustainable and responsible performance by companies? Even without checking the data, we know intuitively from what we see going on in the world that the answer is an unequivocal *no*. With very few exceptions, the global markets today reward the externalisation of social, environmental and ethical costs over the short term.

New York Times journalist and author Thomas Friedman calls this the privatization of benefits and the socialisation of costs, while activist writers like Naomi Klein call it 'the race to the bottom', referring to tendency for companies to locate their production in places with the lowest labour or environmental standards, and hence the lowest costs. To underscore the point, the Vice Fund (VICEX) in the U.S., which only invests in the so-called 'sin' industries like tobacco, alcohol, gambling and armaments (Sanjay 2012) consistently outperforms the market. In fact, over the past 5 years, Vice Fund stocks have gone up 53 %, while the S&P 500 has gone up 12 % and Domini's socially-responsible index has gone up just 5 % (Harris 2012).

However, we do not need to go to extremes to prove the uneconomic nature of responsibility. Why are fairtrade and organic products, or renewable energy, more expensive than more generic products? Why do the banks, oil companies and motor giants remain among the largest and (in some cases) most profitable companies in the world? The fact of the matter is that, beyond basic legal compliance, the markets are designed to serve the financial and economic interests of the powerful, not the idealistic dreams of CSR advocates or the angry demands of civil society activists.

Business leaders seem to agree. In a survey by MIT Sloan Management Review and BCG (2009), more that 70 % of business leaders say their companies have not developed a clear business case for sustainability. Furthermore, the survey of 766 CEOs by Accenture and the UN Global Compact (2010) found that 34 % cited lack of recognition from the financial markets as a barrier to achieving their sustainability goals.

Nestle's Jose Lopez (2010) is candid: 'At the same time that we are coming out with a lot of discussions regarding the importance of sustainability, the market continues. I had hoped that after the world lost 5 trillion dollars in market capitalisation out of this nonsense financial crisis that companies would start to be measured by something else. But the world doesn't seem to be going anywhere other than to measure companies by their market capitalisation.'

References

Accenture and UN Global Compact (2010) A new era in sustainability [CEO survey]

Barnett ML, Salomon RM (2012) Does it pay to be really good? Addressing the shape of the relationship between social and financial performance. Strateg Manag J 33(11):1304–1320

Böhm S, Misoczky MC, Moog S (2012) Greening capitalism? A marxist critique of carbon markets. Organ Stud 33(11):1617–1638

Campbell JT, Eden L, Miller SR (2012) Multinationals and corporate social responsibility in host countries: does distance matter? J Int Bus Stud 43(1):84–106

Carroll AB, Shabana KM (2010) The business case for corporate social responsibility: a review of concepts, research and practice. Int J Manag Rev 12(1):85–105

Chien-Ming C, Delmas MA (2012) Measuring eco-inefficiency: a new frontier approach. Oper Res 60(5):1064–1079

Drucker P (1993) The practice of management. Collins, New York

Economic Policy Institute (2012) The state of working America, 12th edn. Research report

Gallup (2010) Gallup global wellbeing: the behavioral economics of GDP growth. Research report

Griffin JJ, Mahon JF (1997) The corporate social performance and corporate financial performance debate. Bus Soc 36:5

Haanaes K, Michael D, Jurgens J, Rangan S (2013) Making sustainability profitable. Harvard Bus Rev 91(3):110–115

Harris D (2012) Vice funds backer says so-called sin stocks beat market. *ABC News*, 7 October

Helms WS, Oliver C, Webb K (2012) Antecedents of settlement on a new institutional practice: negotiation of the ISO 26000 standard on social responsibility. Acad Manag J 55(5):1120–1145

IPCC (2007) The 4th Assessment Report of the Intergovernmental Panel on Climate Change (IPCC). UNFCCC

Klein N (2000) No logo: no space, no choice, no jobs. Picador

Klein N (2007) The shock doctrine: the rise of disaster capitalism. Metropolitan Books

Laffer AB, Coors A, Winegarden W (2004) Does corporate social responsibility enhance business profitability? Laffer Associates

Lange D, Washburn NT (2012) Understanding attributions of corporate social irresponsibility. Acad Manag Rev 37(2):300–326

Lopez J (2010) Interview with Polly Courtice, Director of the Cambridge Programme for Sustainability Leadership, 17 June 2010

Manifest/MM&K (2012) Remuneration survey 2012

Margolis JD, Walsh JP (2001) People and profits? The search for a link between a company's social and financial performance. Lawrence Erlbaum, Mahwah

MIT Sloan Management Review and BCG (2009) The business of sustainability. Special Report

Ollivier H (2012) Growth, deforestation and the efficiency of the REDD mechanism. J Environ Econ Manag 64(3):312–327
Ortiz I, Cummins M (2011) global inequality: beyond the bottom billion: a rapid review of income distribution in 141 countries. UNICEF report
Sanjay J (2012) Marketing of vice goods: a strategic analysis of the package size decision. Market Sci 31(1):36–51
Stern N (2006) The economics of climate change. CUP, Cambridge
Stiglitz, J., 2010. *Free Markets and the Sinking of the Global Economy*. Penguin
Stolowy H (2012) Creative accounting, fraud and international accounting scandals. Account Rev 87(3):1087–1091
SustainAbility and UNEP (2001) Buried treasure: uncovering the business case for corporate sustainability
SustainAbility, IRC and Ethos Institut (2002) Developing value: the business case for sustainability in emerging markets
Tang Z, Hull CE, Rothenberg S (2012) How corporate social responsibility engagement strategy moderates the CSR-financial performance relationship. J Manag Stud 49(7):1274–1303
Transparency International (2011) Putting corruption out of business survey. Research report
UNDP (2009) Human Development Report 2007/8: Fighting climate change: Human solidarity in a divided world
UNDP (2012) Millennium Development Goals (MDGs) Report. United Nations
Vogel D (2005) The market for virtue: the potential and limits of corporate social responsibility. Brookings Institution Press
Walls JL, Berrone P, Phan PH (2012) Corporate governance and environmental performance: is there really a link?. Strateg Manag J 33(8):885–913
Whiteman G, Walker B, Perego P (2013) Planetary boundaries: ecological foundations for corporate sustainability. J Manag Stud 50(2):307–336
Yongtae K, Seok PM, Wier B (2012) Is earnings quality associated with corporate social responsibility? Account Rev 87(3):761–796

Chapter 4
CSR 2.0 as a New Metaphor

Abstract The metaphor of 1.0 and 2.0 is explained as an appropriate analogy for the changes needed in CSR, drawing parallels with the evolution of Web 1.0 to Web 2.0. Key shifts described include the change from small scale, one-way-communication and standardised approaches to mass scale, interactive, user-generated solutions.

Keywords Web 2.0 · CSR 1.0 · CSR 2.0 · Long tail · Openness · Peering · Sharing · Global · Mass customization · Wikinomics

4.1 Introduction to Web 2.0

Throughout my 20-year career in corporate sustainability and responsibility, Ray Anderson and Anita Roddick are the kinds of pioneers I have looked to for hope and inspiration. The frustration has been that these 'radical industrialists' have always remained the exception, rather than the rule. They are the outliers, which is fine if—in line with Everett Rogers' (1962) Diffusion of Innovation model—they are the innovators that make up 2.5 % of the population. The problem is that most of their ideas and practices have not diffused to the early adopters and the early majority, let alone the late majority and laggards.

So what will it take to get the kind of transformation we need to move beyond innovation towards mass change? I find an analogy is always helpful and in early 2008, I discovered the perfect metaphor: Web 2.0. The term, of course, had been around for a while—coined by IT consultant Darcy DiNucci (1999) in an article called 'Fragmented Future' and popularised in 2004 by the landmark O'Reilly Media Web 2.0 conference. O'Reilly's (2005) article 'What is Web 2.0' had already become an early adopters' touchstone for a rapidly evolving new lexicon, and remains a classic piece. People like me, part of the technosphere's early majority, were a bit slower in waking up, and it took Tapscott and Williams' (2006) book *Wikinomics* to switch me on to the revolution in progress.

W. Visser, *CSR 2.0*, SpringerBriefs in Organisational Studies,
DOI: 10.1007/978-3-642-40874-8_4, © The Author(s) 2014

Before coming to why Web 2.0 is a good metaphor for the transformation of CSR, let me try to bed down the concept. Today, Wikipedia defines Web 2.0 as 'web applications that facilitate interactive information sharing, inter-operability, user-centered design and collaboration.' Fair enough, but let's dig a little deeper, drawing on the term's evolution. In 1999, DiNucci was writing for programmers, challenging them to adapt to the increasing use of portable Web-ready devices.

This was just a small part of what Web 2.0 would come to mean. In 2005, O'Reilly brainstormed a far more wide ranging list of examples and contrasts between Web 1.0 and Web 2.0. Examples included DoubleClick versus Google AdSense, Britannica Online versus Wikipedia, personal websites versus blogging, publishing versus participation, directories (taxonomy) versus tagging (folksonomy) and stickiness versus syndication, to mention but a few. His article concluded with seven core competencies of Web 2.0 companies:

1. Services, not packaged software, with cost-effective scalability;
2. Control over unique, hard-to-recreate data sources that get richer as more people use them;
3. Trusting users as co-developers;
4. Harnessing collective intelligence;
5. Leveraging the long tail through customer self-service;
6. Software above the level of a single device; and
7. Lightweight user interfaces, development models and business models.

Tapscott and Williams (2006) gave an applied view on Web 2.0 in the form of 'wikinomics', which they defined as 'the effects of extensive collaboration and user-participation on the marketplace and corporate world'. Wikinomics, they said, is based on four principles:

1. *Openness*, which includes not only open standards and content but also financial transparency and an open attitude towards external ideas and resources;
2. *Peering*, which replaces hierarchical models with a more collaborative forum, for which the Linux operating system is a quintessential example;
3. *Sharing*, which is a less proprietary approach to (among other things) products, intellectual property, bandwidth and scientific knowledge; and
4. *Acting globally*, which involves embracing globalisation and ignoring physical and geographical boundaries at both the corporate and individual level.

4.2 Application to CSR 2.0

By May 2008, it was clear to me that this evolutionary concept of Web 2.0 held many lessons for CSR. I published my initial thoughts in a short article online entitled 'CSR 2.0: The New Era of Corporate Sustainability and Responsibility' (Visser 2008c), in which I said:

Table 4.1 Similarities between Web 1.0 and CSR 1.0

Web 1.0	CSR 1.0
A flat world just beginning to connect itself and finding a new medium to push out information and plug advertising	A vehicle for companies to establish relationships with communities, channel philanthropic contributions and manage their image
Saw the rise to prominence of innovators like Netscape, but these were quickly out-muscled by giants like Microsoft with its Internet Explorer	Included many start-up pioneers like Traidcraft, but has ultimately turned into a product for large multinationals like Wal-Mart
Focused largely on the standardised hardware and software of the PC as its delivery platform, rather than multi-level applications	Travelled down the road of 'one size fits all' standardisation, through codes, standards and guidelines to shape its offering

The field of what is variously known as CSR, sustainability, corporate citizenship and business ethics is ushering in a new era in the relationship between business and society. Simply put, we are shifting from the old concept of CSR—the classic notion of 'Corporate Social Responsibility', which I call CSR 1.0—to a new, integrated conception—CSR 2.0, which can be more accurately labelled 'Corporate Sustainability and Responsibility'. The allusion to Web 1.0 and Web 2.0 is no coincidence. The transformation of the internet through the emergence of social media networks, user-generated content and open source approaches is a fitting metaphor for the changes business is experiencing as it begins to redefine its role in society. Let's look at some of the similarities (Table 4.1).

As our world becomes more connected and global challenges like climate change and poverty loom ever larger, businesses that still practice CSR 1.0 will (like their Web 1.0 counterparts) be rapidly left behind. Highly conscientised and networked stakeholders will expose them and gradually withdraw their social licence to operate (Scherer et al. 2013) (Table 4.2).

By contrast, companies that embrace the CSR 2.0 era will be those that collaboratively find innovative ways tackle our global challenges and be rewarded in the marketplace as a result.

4.3 Long Tail of CSR 2.0

While exploring these ideas, I read Chris Anderson's (2008) Web 2.0 book *The Long Tail* and wondered whether it could also be applied to CSR. The long tail—named after the extended tail of a statistical distribution curve—is the idea that selling less to more people is big business. It's the business model that has spawned the most successful companies of the Web 2.0 age. The long tail questions the conventional wisdom that says success is about generating 'blockbusters' and 'superstars'—those rare few products and services that become runaway bestsellers.

Table 4.2 Similarities between Web 2.0 and CSR 2.0

Web 2.0	CSR 2.0
Being defined by watchwords like 'collective intelligence', 'collaborative networks' and 'user participation'	Being defined by 'global commons', 'innovative partnerships' and 'stakeholder involvement'
Tools include social media, knowledge syndication and beta testing	Mechanisms include diverse stakeholder panels, real-time transparent reporting and new-wave social entrepreneurship
Is as much a state of being as a technical advance—it is a new philosophy or way of seeing the world differently	Is recognising a shift in power from centralised to decentralised; a change in scale from few and big to many and small; and a change in application from single and exclusive to multiple and shared

So is there a long tail of CSR? And if so, what does it look like? To me, the long tail of CSR is all about extending the reach of CSR, and improving its ability to satisfy specific social and environmental needs. Let's use Anderson's enablers as a framework for thinking about this.

Democratising the tools of CSR production

This is about breaking CSR silos and extending CSR beyond multinationals. At the early stages of CSR adoption, it is often confined to Public Relations, Corporate Affairs or Marketing departments. As CSR implementation matures, responsibility tends to migrate to specialised CSR departments of various descriptions (environment, health and safety, accountability, corporate citizenship, etc.). However, these versions of CSR are like the Hollywood model of blockbuster films. They suggest that CSR is about a few, high visibility programmes that are designed by CSR experts and delivered by big companies.

By contrast, democratising CSR production would mean firstly embedding CSR across the organisation—making it the responsibility of operations managers, financial managers, shop floor workers, basically everyone. This is only possible if CSR becomes part of the culture and incentive systems of an organisation. CSR would also need to be extended beyond the usual suspects (i.e. the high profile, branded multinationals) to the less visible B2B (business to business) and national (rather than multinational) organisations, as well as to SMEs (small and medium sized enterprises) and down the supply chain.

Democratising the tools of CSR distribution

To date, CSR has mainly been 'distributed' via a few select projects—typically philanthropic or charitable activities—in which the company offers its help to the 'less fortunate masses'. Usually, the nature and scope of CSR activities is determined top-down and offered as a fairly undifferentiated 'service', e.g. Nike might decide to focus on sponsoring sports teams, events and celebrities and Coca Cola might choose water as its key CSR issue. The most common delivery mechanisms are money (sponsorship and other forms of charity), or for the more advanced companies, adhering to generic CSR codes and standards.

By contrast, democratising the tools of CSR distribution should include allowing staff to participate in CSR delivery through volunteer programmes, and developing more geographically tailored and sector-specific CSR codes and standards, such as the Roundtable on Sustainable Palm Oil, or the Global Reporting Initiative guidelines for HIV/AIDS reporting. Beyond this, embracing Bottom of the Pyramid (BOP) markets (Simanis 2012) and supporting social entrepreneurs will allow the reach of CSR to be extended so that the needs of formerly unserved or underserved people can be met (Mair et al. 2012).

Connecting CSR Supply and Demand

Traditionally, CSR has been offered in the form of grants by multinational head-offices, who control the budget and set the criteria by which prospective philanthropic projects should be selected. For the more advanced companies, this has been extended to adherence by their operations to corporate codes of CSR practice and communicating this through CSR reports (Davidson and Stevens 2013). Demand has typically come from community groups applying to corporate foundations for funding, or NGOs taking an activist approach to demanding improved CSR practices (Heyes and Kapur 2012).

By contrast, connecting the long tail of CSR supply and demand will rely increasingly on cross-sector partnerships and multi-stakeholder groups (Koschmann et al. 2012). For example, Rio Tinto works with the World Conservation Union to identify biodiversity needs and satisfy them through appropriate CSR activities. Companies may also use extended stakeholder networks of community groups, social entrepreneurs and microcredit enterprises (Bauer et al. 2012) to better match their capacity to make a positive impact among those who can most benefit, as BP is doing with smokeless stoves in India and SC Johnson is doing with cleaning products in Kenya.

Hence, applying the long tail concept to CSR requires a different way of thinking about how CSR is generated, delivered and managed. It means making CSR a more inclusive and embedded process within the company, and a more diverse and far-reaching set of activities outside the company. It also means creating meaningful stakeholder partnerships to ensure that the right kinds of CSR benefit the right groups of people, where and when they need it (Hillenbrand et al. 2013). The long tail in a nutshell, according to Anderson, is: ‘culture unfiltered by scarcity’. By extension, the long tail of CSR in a nutshell is: ‘responsibility liberated by collaboration’.

4.4 Principles of CSR 2.0

Initial responses to my framing of CSR 2.0 were largely positive and confirmed that I was onto something—perhaps a new language or conceptualisation of responsibility, or at the very least a nexus for talking about the radical changes needed in CSR. However, I felt it needed an institutional vehicle if it was going to have any chance of success, and so CSR International was born, with the express mission to be an incubator for CSR 2.0. The think tank was launched on 3 March

2010 in London, complete with the ritualistic burial of the old CSR and its rebirth as CSR 2.0.

It quickly became clear, however, that a metaphor can only take us so far. What was needed was a set of principles against which we could test CSR. These went through a few iterations, but I eventually settled on five, which form a kind of mnemonic for CSR 2.0: Creativity (C), Scalability (S), Responsiveness (R), Glocality (2) and Circularity (0). These principles, which will be explored in detail in a later section, can be described briefly as follows:

Creativity (C)—The problem with the current obsession with CSR codes and standards (including the new ISO 26000 standard) is that it encourages a tick-box approach to CSR. But our social and environmental problems are complex and intractable. They need creative solutions, like Freeplay Energy's wind-up technology or Vodafone's M-Pesa money transfer scheme (Visser 2013).

Scalability (S)—The CSR literature is liberally sprinkled with charming case studies of truly responsible and sustainable projects. The problem is that so few of them ever go to scale. We need more examples like Wal-Mart 'choice editing' by converting to sustainable fish, Tata creating the affordable eco-efficient Nano car or Muhammad Yunus's Grameen microfinance model (Beard 2012).

Responsiveness (R)—More cross-sector partnerships and stakeholder-driven approaches are needed at every level, as well as more uncomfortable, transformative responsiveness, which questions whether particular industries, or the business model itself, are part of the solution or part of the problem. A good example of responsiveness is the Corporate Leaders Group on Climate Change (Adey and Visser 2007).

Glocality (2)—This means 'think global, act local'. In a complex, interconnected, globalising world, companies (and their critics) will have to become far more sophisticated in combining international norms with local contexts, finding local solutions that are culturally appropriate, without forsaking universal principles. We are moving from an 'either-or' one-size-fits-all world to a 'both-and' strength-in-diversity world.

Circularity (0)—Our global economic and commercial system is based on a fundamentally flawed design, which acts as if there are no limits on resource consumption or waste disposal. Instead, we need a cradle-to-cradle approach, closing the loop on production and designing products and processes to be inherently 'good', rather than 'less bad', as Shaw Carpets does.

4.5 Shifting from CSR 1.0 to CSR 2.0

These principles are the acid test for future CSR practices. If they are applied, what kind of shifts will we see? In my view, the shifts will happen at two levels. At a meta-level, there will be a change in CSR's ontological assumptions or ways of seeing the world. At a micro-level, there will be a change in CSR's methodological practices or ways of being in the world.

The meta-level changes can be described as follows: Paternalistic relationships between companies and the community based on philanthropy will give way to more equal partnerships (Rufín and Rivera-Santos 2012). Defensive, minimalist responses to social and environmental issues will be replaced by proactive strategies and investment in growing responsibility markets, such as clean technology (Eesley and Hannah 2012). Reputation-conscious public-relations approaches to CSR will no longer be credible and so companies will be judged on actual social, environmental and ethical performance, i.e. are things getting better on the ground in absolute, cumulative terms? (Cho et al. 2012).

Although CSR specialists still have a role to play, each dimension of CSR 2.0 performance will be embedded and integrated into the core operations of companies. Standardised approaches will remain useful as guides to consensus, but CSR will find diversified expression and implementation at very local levels. CSR solutions, including responsible products and services, will go from niche 'nice-to-haves' to mass-market 'must-haves'. And the whole concept of CSR will lose its Western conceptual and operational dominance, giving way to a more culturally diverse and internationally applied concept (Table 4.3).

Table 4.3 Meta-level Ontological Shifts

CSR 1.0	CSR 2.0
Philanthropic	Collaborative
Risk-based	Reward-based
Image-driven	Performance-driven
Specialised	Integrated
Standardised	Diversified
Marginal	Scalable
Western	Global

How might these shifting principles manifest as CSR practices? Supporting these meta-level changes, the anticipated micro-level changes can be described as follows: CSR will no longer manifest as luxury products and services (as with current green and fairtrade options), but as affordable solutions for those who most need quality of life improvements (Ying-Ching and Chang 2012). Investment in self-sustaining social enterprises will be favoured over cheque-book charity (Desa 2012; Bugg-Levine et al. 2012). CSR indexes, which rank the same large companies over and over (often revealing contradictions between indexes) will make way for CSR rating systems, which turn social, environmental, ethical and economic performance into corporate scores (A+, B−, etc., not dissimilar to credit ratings) and which analysts and others can usefully employ in their decision making.

Reliance on CSR departments will disappear or disperse, as performance across responsibility and sustainability dimensions are increasingly built into corporate performance appraisal and market incentive systems. Self-selecting ethical consumers will become irrelevant, as CSR 2.0 companies begin to choice-edit, i.e. cease offering implicitly 'less ethical' product ranges, thus allowing guilt-free shopping. Post-use liability for products will become obsolete, as the service-lease and take-back economy goes mainstream (Agrawal et al. 2012). Annual CSR reporting will be replaced by online, real-time CSR performance data flows. Feeding into these live communications will be Web 2.0 connected social networks that allow 'crowdsourcing', instead of periodic meetings with rather cumbersome stakeholder panels. And typical CSR 1.0 management systems standards like ISO 14001 will be less credible than new performance standards, such as those emerging in climate change that set absolute limits and thresholds (Table 4.4).

4.6 Web 2.0 as a Transformational Tool

My main intention with CSR 2.0 is to apply Web 2.0 as a metaphor. However, recently I have also been thinking about a more literal interpretation of CSR 2.0, namely how the tools of Web 2.0 can be used as a transformational force in CSR. If we take Tapscott and Williams' (2006) four principles of Web 2.0 (openness, peering, sharing and acting globally), plus the essential principle derived from Anderson's (2008) 'long tail' concept (mass customization), this gives us a kaleidoscopic lens through which to look at the future of business, as summarised in Table 4.5 and described briefly.

Table 4.4 Micro-level Methodological Shifts

CSR 1.0	CSR 2.0
CSR premium	Base of the pyramid
Charity projects	Social enterprise
CSR indexes	CSR ratings
CSR departments	CSR incentives
Product liability	Choice editing
Ethical consumerism	Service agreements
CSR reporting cycles	CSR data streams
Stakeholder groups	Social networks
Process standards	Performance standards

4.6.1 Principle 1: Openness

4.6.1.1 Practice 1: Net Value Footprinting

Business has evolved over the past two decades from being highly opaque to gradually embracing more transparent disclosure practices (Dhaliwal et al. 2012). This has been a result of regulation (such as the Toxic Release Inventory in the U.S., which requires thousands of American companies to report over 650 toxic chemicals) and voluntary efforts (such as the GRI, which now has the fourth iteration of is Sustainability Reporting Guidelines).

Table 4.5 Web 2.0 Practices for CSR 2.0

Principles	Practices
Openness	Net value footprinting
	Forensic impact analysis
Peering	Stakeholder crowdsourcing
	Disruptive partnerships
Sharing	Open-sourcing
	Wiki-ratings
Acting globally	Prototyping
	Smart mobbing
Mass customization	App farming
	Plug-and-play

In a Web 2.0 world, however, transparency requirements are taken to another level. Companies are expected to go beyond GRI-based reporting, to measure and disclose their impacts across the entire product life cycle or value chain. This process of quantifying business's economic, social and environmental costs to society is sometimes called full cost accounting, or internalizing externalities. I call it Net Value Footprinting.

Net Value Footprinting is being pioneered by the likes of Patagonia (with their Footprint Chronicles™), Puma (with their Environmental Profit and Loss statement), The Economics of Ecosystems and Biodiversity (TEEB) study, and the Global Footprint Network.

4.6.1.2 Practice 2: Forensic Impact Analysis

While progressive companies are steadily improving their transparency, there will also be millions of irresponsible companies that try to fly under the radar of regulation and public scrutiny. In an effort to be lowest cost producers or preferred suppliers to big brand multinationals, they will deliberately externalize social and environmental costs by running polluting operations that exploit cheap labour and abuse human rights (Crane 2013).

But in a Web 2.0 world, these rogue businesses will be caught and exposed through the emerging practice of what I call Forensic Impact Analysis. This will happen through a combination of traceability technology (which finds the electronic footprints left by all businesses in the supply chain), forensic substance analysis (which can identify the source of fibres, chemicals and other product components) and vigilant activists and consumers (who will capture malpractices on mobile phones using photographs, videos and audio recordings, and leak these via online social media).

Forensic Impact Analysis is being pioneered by the food industry, which uses barcodes or RFID tags and other tracking media to monitor every step of their production process (GrapeNet in India is an example). Other examples include Karmayog (which allows online whistleblowing on corruption in India) and Wikileaks (which exposed Trafigura's dumping of toxic waste along the Ivory Coast).

4.6.2 Principle 2: Peering

4.6.2.1 Practice 3: Stakeholder Crowdsourcing

Companies from the Web 1.0 era still believe that focus groups, public meetings, stakeholder panels and the occasional online or in-store survey are adequate for taking the pulse of their stakeholders. At the same time, they are generally distrustful of ideas or solutions from outside their organisations. In short, they suffer from the 'not invented here' syndrome.

By contrast, Web 2.0 savvy companies realise that the world has moved into an era of crowdsourcing—a term coined by Jeff Howe in 2006 and closely linked to the earlier idea of 'wisdom of crowds' popularized by James Surowiecki. Turning this concept into practice, future business will increasingly use filtered, expert 'crowds' to monitor their reputation, get feedback on sustainable products innovations and solicit help in solving difficult ethical dilemmas (Healy and Ramanna 2013).

Stakeholder Crowdsourcing is being pioneered by companies like Sony, through its two online campaigns, Open Planet Ideas and FutureScapes (to generate new sustainable technology ideas) and platforms like OpenEyeWorld, which General Electric has used to crowdsource feedback on its sustainability communications.

4.6.2.2 Practice 4: Disruptive Partnerships

Companies have had a decade to get used to the idea of cross-sector partnerships (Koschmann et al. 2012), which have been heavily promoted through the United Nations and given a boost through inclusion in the Millennium Development Goals and being spotlighted at the World Summit on Sustainable Development in Johannesburg in 2002.

In a Web 2.0 world, however, business is expected to get into more challenging partnerships—collaborations which disrupt the status quo. For example, Greenpeace very effectively used social media to campaign against Nestle's Kit-Kat brand, after finding an Indonesian supplier that was clearing tropical rainforest to grow palm oil. A year later, Greenpeace praised Nestle for their No Deforestation commitment through its challenging partnership with TFT, a sustainable forestry NGO.

Disruptive Partnerships are being pioneered by the likes of Rio Tinto (partnering with the World Conservation Union to reduce their biodiversity impacts), BASF (through their Strategic Alliance for the Fortification of Oil and Other Staple Foods partnership with GIZ), and Netherlands flooring company Desso (using their Circle of Architects creative forum).

4.6.3 Principle 3: Sharing

4.6.3.1 Practise 5: Open-Sourcing

One of the biggest changes in the society over the past 10 years has been the explosion of social media. But this revolution goes beyond sharing our holiday photos on Facebook or the micro-blogging the minutiae of our lives on Twitter. The more fundamental innovation is a shift in thinking and practice towards 'open-sourcing', which at its heart is about co-creation.

Let's look at an example from the pharmaceutical industry to illustrate the point. After a decade under siege—with Big Pharma being accused of overpricing their patented brands and blocking access to cheaper generic (and often life saving) drugs—GlaxoSmithKline (GSK)'s CEO Andrew Witty committed GSK to put any chemicals or processes over which it has intellectual property rights that are relevant to finding drugs for neglected diseases into a 'patent pool', so they can be explored by other researchers.

Other pioneering examples include the World Business Council for Sustainable Development's (WBCSD) Eco-Patent Commons and the Creative Commons' GreenXchange, both of which allow companies to share their intellectual property 'for the common good', especially on issues like waste, pollution, climate change and energy.

4.6.3.2 Practise 6: Wiki-Ratings

Another feature of Web 2.0 design is that it easily allows users to express an opinion on others' content—from the ubiquitous thumbs-up 'Like' feature on Facebook, to the fresh-red versus rotten-green tomato movie rating system on rottentomatoes.com.

Now, we are going beyond these simplistic approaches to dynamic, wiki-based platforms that allow the public to rate—and comment in detail—on the economic, governance, social and environmental performance of companies. One such innovative platform, where I serve on the advisory board, is Wikirate, developed by Philipp Hirche. Not only does Wikirate use a crowdsourcing approach to ratings, but in much the same way as Wikipedia, it allows for real-time updating. Hence, an ethical infringement, or a sustainability innovation, will be reflected almost immediately in the company's wikirating.

Other pioneering examples in the ratings space are GoodGuide, WeGreen, Project Label and Scryve, although judging by SustainAbility's 'Rate the Raters' analysis, none of the 108 rating systems identified employ a methodology quite so democratic and transparent as Wikirate.

4.6.4 Principle 4: Acting Globally

4.6.4.1 Practise 7: Prototyping

Innovation has always used prototyping—i.e. designing a working sample of new products and services. The difference in a Web 2.0 world is that prototypes are launched early, as imperfect versions, to solicit rapid user feedback in a process often called 'beta-testing'.

One way to bring about such rapid, open-source prototyping is through competitions. Take the X-Prize, for example, which describes its mission as 'bringing about radical breakthroughs for the benefit of humanity' in five areas: education; global development; energy and environment; life sciences; and exploration. Through this platform, multimillion-dollar prizes are offered for innovative solutions in everything from 'progressive automotive' and 'oil cleanup' to 'health sensors' and 'diagnostic technologies'.

Another pioneering example is Virgin's $25 million Earth Challenge, for 'a commercially viable design which results in the net removal of anthropogenic, atmospheric greenhouse gases so as to contribute materially to the stability of the Earth's climate system.'

4.6.4.2 Practise 8: Smart Mobbing

Web 2.0 technologies have spawned a new type of protest activity, called smart mobbing. This simply means using real-time media and sharing platforms—especially SMS texts and status updates (like tweets on Twitter)—to rapidly organise a crowd.

Examples include 'viral' text messaging in the Philippines that helped to oust former President Joseph Estrada in 2001 and the use of Twitter during the Arab Spring uprisings in 2011. Smart mobs can also co-ordinate virtual activity, such as when the 'hacktivist' group Anonymous encouraged its followers to launch cyber attacks against Visa, MasterCard, PayPal and other companies opposing Wikileaks in 2011. Similarly, Greenpeace encouraged smart mobbing following its 2010 campaign against Nestle's Kit-Kat brand. The campaign video was viewed by half a million people in 4 days, and unleashed a flood of angry comments on Nestle's Facebook page.

Smart mobbing can also be used positively, such as when 'Mission 4636' created an SMS text mapping emergency communications system after the 2010 Haiti earthquake. In future, companies and governments will increasingly need to anticipate and respond to activist smart mobs, as well as seeding their own.

4.6.5 Principle 5: Mass Customization

4.6.5.1 Practise 9: App Farming

Despite some great new gadgets over the past few years—such as the iPad—the war of the computing giants has turned into a 'battle of apps'. Underlying this explosive trend, by April 2012, Apple claimed to have 615,000 apps and 43,000 iOS-based app developers, while Google had 430,000 apps and 10,000 Android-based app developers.

Apps (software applications) are essentially neatly packaged, user-friendly online services, ranging from games (e.g. Angry Birds, Scrabble) and music (e.g. Spotify, Shazam) to education (e.g. NASA, Spelling Bee) and business (e.g. HBR Tips, EasyMoney 1.0). There is also a new generation of apps focused on social and environmental solutions. Google Play lists more than 400 sustainability-related apps. The most popular is BlaBlaCar, which connects drivers with empty seats with people looking for a ride, allowing users to post on, and search, the biggest European car sharing community.

Other popular apps in this genre include GoodGuide (for ethical shopping), carbon footprint calculators (Google Play lists five) and educational games like 'Sustainable me'. Hence, businesses of the future will be judged on whether they can seed and grow farms of apps that provide solutions to the world's most serious challenges.

4.6.5.2 Practise 10: Plug-and-Play

The final Web 2.0 savvy practise is to think in terms of 'plug-and-play' solutions. Essentially, this is a form of smart technology that detects its operating environment, installs whatever software is needed and is operational without any action by the user.

To take a simple example, rather than having to manually unplug or switch off household electrical devices to save energy, a plug-and-play device in the home automatically detects all idle devices and disables them remotely. Similar approaches apply to optimal energy-efficient heating and cooling of buildings (i.e. indoor climate regulation) and low-carbon driving, which automatically chooses the emission-minimizing acceleration and cruise speeds.

Plug-and-play also applies to our shopping preferences. In the future, we will have automatic product filters that match our personal preferences—whether it is for fairtrade, organic, beauty without cruelty, or health. When shopping online, we will only see products that match our personal criteria. Similarly, in-store we will be alerted to products that meet our standards—a process achieved through auto-scanning by our mobile devices of in-store barcodes and associated criteria-linked product databases.

4.7 Conclusion

The message is clear for business. Web 2.0 is not just about everybody being continuously online. Rather, it is about a new business mind-set that thinks in terms of the collective intelligence of its stakeholders, the co-creation of solutions to our global challenges, and the use of technology to achieve speed and scale in spreading innovation to those parts of the world that have the biggest and most urgent unmet needs (Khavul and Bruton 2013; George et al. 2012).

References

Adey M, Visser W (2007) A new model of business-government policy dialogue on sustainability: the case of the corporate leaders group on climate change. Cambridge Programme for Industry Research Paper Series 3, University of Cambridge
Agrawal VV, Ferguson M, Toktay LB, Thomas VM (2012) Is leasing greener than selling? Manag Sci 58(3):523–533
Anderson C (2008) The long tail: why the future of business is selling less of more. 2nd edn. Hyperion
Bauer M, Chytilov J, Morduch J (2012) Behavioral foundations of microcredit: experimental and survey evidence from rural India. Am Econ Rev 102(2):1118–1139
Beard A (2012) Muhammad Yunus. Harvard Bus Rev 90(12):136
Bugg-Levine A, Kogut B, Kulatilaka N (2012) A new approach to funding social enterprises. Harvard Bus Rev 90(1–2)
Cho CH, Guidry RP, Hageman AM, Patten DM (2012) Do actions speak louder than words? An empirical investigation of corporate environmental reputation. Acc Organ Soc 37(1):14–25
Crane A (2013) Modern slavery as a management practice: exploring the conditions and capabilities for human exploitation. Acad Manag Rev 38:49–69
Davidson BI, Stevens DE (2013) Can a code of ethics improve manager behavior and investor confidence? An Experimental Study. Account Rev 88(1):51–74
Desa G (2012) Resource mobilization in international social entrepreneurship: bricolage as a mechanism of institutional transformation. Entrepr Theory Pract 36(4):727–751
Dhaliwal DS, Radhakrishnan S, Tsang A, George YY (2012) Nonfinancial disclosure and analyst forecast accuracy: international evidence on corporate social responsibility disclosure. Account Rev 87(3):723–759
DiNucci D (1999) Fragmented future. Print magazine, pp 32, 221–222
Eesley CE, Hannah D (2012) Ion B. Vasi: winds of change: the environmental movement and the global development of the wind energy industry. Adm Sci Q 57(2):359–362
George G, McGahan AM, Prabhu J (2012) Innovation for inclusive growth: towards a theoretical framework and a research agenda. J Manag Stud 49(4):661–683
Healy PM, Ramanna K (2013) When the crowd fights corruption. Harvard Bus Rev 91(1):122–128
Heyes A, Kapur S (2012) Community pressure for green behavior. J Environ Econ Manag 64(3):427–441
Hillenbrand C, Money K, Ghobadian A (2013) Unpacking the mechanism by which corporate responsibility impacts stakeholder relationships. Br J Manag 24(1):127–146
Khavul S, Bruton GD (2013) Harnessing innovation for change: sustainability and poverty in developing countries. J Manag Stud 50(2):285–306

Koschmann MA, Kuhn TR, Pfarrer MD (2012) A communicative framework of value in cross-sector partnerships. Acad Manag Rev 37(3):332–354

Mair J, Marti I, Ventresca MJ (2012) Building inclusive markets in rural bangladesh: how intermediaries work institutional voids. Acad Manag J 55(4):819–850

O'Reilly T (2005) What is Web 2.0? Design Patterns and Business Models for the Next Generation of Software. Article on http://oreilly.com [Accessed on 28 March 2013]

Rogers EM (1962) Diffusion of innovations. Free Press, New York

Rufín C, Rivera-Santos M (2012) Between commonweal and competition: understanding the governance of public–private partnerships. J Manag 38(5):1634–1654

Scherer AG, Palazzo G, Seidl D (2013) Managing legitimacy in complex and heterogeneous environments: sustainable development in a globalized world. J Manag Stud 50(2):259–284

Simanis E (2012) Reality check at the bottom of the pyramid. Harvard Bus Rev 90(6):120–125

Tapscott D, Williams AD (2006) Wikinomics: how mass collaboration changes everything. Portfolio

Visser W (2008c) CSR 2.0: the new era of corporate sustainability and responsibility. CSR International Inspiration Series, No. 1

Visser W (2013) Corporate sustainability & responsibility: an introductory text on CSR theory & practice—past, present & future. Kaleidoscope Futures, London

Ying-Ching L, Chang CA (2012) Double standard: the role of environmental consciousness in green product usage. J Market 76(5):125–134

Chapter 5
The Principles of CSR 2.0

Abstract The emergent CSR 2.0 practice is described in detail by elaborating on five principles underlying this new approach, including: creativity, scalability, responsiveness, glocality (global–local balance) and circularity (closed-loop production). Each principle is explained conceptually, as well as illustrated with progressive case studies.

Keywords Creativity · Innovation · Scaling · Mainstreaming · Partnership · Stakeholders · Glocality · Context · Circular economy · Cradle to cradle

There are five principles that are the acid test for CSR 2.0: creativity, scalability, responsiveness, glocality and circularity. These are summarised in Table 5.1 below and then described in more detail in the sections that follow.

5.1 Creativity

One of the great dangers of the age of management—especially its focus on standardisation—is that it does not foster the kind of creativity that is needed to solve the complex social, environmental and ethical problems that we face. The reasons are fairly obvious. An incremental approach, in which companies voluntarily set their own CSR-related objectives, does not tend to create the kind of stretch targets that incubate innovation. Also, standardisation is by its very nature a compliance-based approach, with systems, procedures, measures and audits. As a result, those running on the standards treadmill develop a tick-box mentality, rather than thinking outside the box.

This was somewhat evident in a McKinsey (2010) survey, which showed that while more than 50 % of executives considered sustainability 'very' or 'extremely' important in a wide range of areas—including new-product development, reputation building, and overall corporate strategy—only around 30 % said their

W. Visser, *CSR 2.0*, SpringerBriefs in Organisational Studies,
DOI: 10.1007/978-3-642-40874-8_5, © The Author(s) 2014

Table 5.1 Principles of CSR 2.0

Principle	Description	Examples
Creativity	Moving beyond 'tick-box' CSR approaches like codes & standards to embrace social & environmental innovation & entrepreneurship, creating products & services to solve sustainability challenges	Free-play's 'wind-up' energy technologies (radios, torches, laptops); Vodafone's M-Pesa scheme for money transfer by text
Scalability	Moving beyond 'ethical consumerism' CSR approaches like fairtrade & eco-labelling to embrace 'choice editing' where only sustainable & responsible options are offered for entire product or service lines	Walmart's Marine Stewardship Council (MSC) certified fish product lines; Co-operative Bank's ethical banking & insurance services
Responsiveness	Moving beyond 'shareholder-driven' CSR approaches like charity donations & promotional campaigns to embrace stakeholder responsiveness, including impact investing & stakeholder feedback mechanisms	The Corporate Leaders Group on Climate Change; the Carbon Disclosure Project's climate solutions investment strategy
Glocality	Moving beyond imperial and parochial CSR approaches to embrace 'think global, act local' practices, which follow international guidelines, but tailor context-specific solutions	BHP-Billiton's dual strategies for climate change (global) & malaria prevention (local); Local Agenda 21 actions, based on global principles
Circularity	Moving beyond 'end of pipe' CSR approaches to embrace 'cradle to cradle' practices, which close the loop on production & service processes, with the goal of zero waste, zero toxics & 100 % renewable resources	Fuji-Xerox's product take-back scheme in Asia; Shaw Carpets' closed loop production, end-of-life collection & recycling system

companies actively seek opportunities to invest in sustainability or embed it in their business practices. Not that corporate sustainability and responsibility lacks the potential for innovation (Xueming and Shuili 2012). Hart (1997) comments that 'greening has been framed in terms of risk reduction, reengineering, or cost cutting. Rarely is greening linked to strategy or technology development, and as a result, most companies fail to recognise opportunities of potentially staggering proportions' (see also Sengupta 2012).

More than 10 years later, these sentiments were echoed by Nidumolu et al. (2009) who observed that 'companies won't innovate successfully—and as a result won't grow—unless they throw themselves whole hog into green initiatives.' Their conclusion was that 'smart companies now treat sustainability as innovation's new frontier', a finding echoed by Esty and Charnovitz (2012). Similarly, Hollender and Green (2010) declare that 'green marketing campaigns don't cut it anymore; insurgent good companies focus on innovation rather than reputation'.

An industry that is proving to be an effective agent of the kind of 'creative destruction' we need is the mobile phones sector. For instance, Vodafone, beyond simply expanding traditional mobile services to Africa, have also used the continent as an incubator for innovation. In Kenya in 2005, for example, 80 % of the population did not have a bank account. Also, more money was coming into the country through remittances from family members living abroad than through overseas development assistance. However, these transfers were expensive, with Western Union typically taking a big slice in commission. Hence, Vodafone developed and piloted a new service called M-Pesa, whereby customers could use their mobile phones to perform basic financial services, including depositing, withdrawing and transferring money using SMS texts (Visser 2013).

For many, the service has been life-changing, giving access to financial services from which they were excluded and allowing them to receive remittance payments from the UK directly. Besides employing and empowering thousands of M-Pesa agents, the scheme has also cut out a lot of corruption, as all transactions are electronic. The M-Pesa service has been extended to Tanzania, Afghanistan and South Africa. In 2009, Safaricom also launched the continent's first commercial solar powered mobile phone, the Coral-200. Building on the success of Vodafone and others, a 2010 study by Arthur D. Little estimates that global transaction volume in mobile financial services will reach approximately $280 billion by 2015.

Anurag Gupta, founder of A Little World in India, is a social entrepreneur who has extended mobile banking even further (Arora and Cummings 2010). Using a combination of mobile phones (for voice identification and storage of 50,000 customer records) and an integrated printer-biometric scanner (for receipts and fingerprints), he has created 'tiny branches' in India's rural villages that are run by one person, cost $80 a month and make basic banking services accessible to millions that were previously excluded (Visser 2011).

Gupta sees the branch network as an enabler to deliver all kinds of other essential services to India's rural poor. Already, he has innovated rechargeable

LED light boxes to replace polluting and hazardous kerosene lamps, as well as enhancements to wood or cow-dung burning stoves, using a fan that halves cooking time, halves fuel requirements and almost eliminates the poisonous smoke. Future innovations include water filters, bicycles, televisions, spectacles, radios, medicines and textbooks.

To make all these products affordable, Gupta plans to use a lease-purchase model, whereby costs are divided into weekly instalments for 6, 12, 18 or 24 months, depending on the product. So, for example, a rural villager pays just a few Rupees for 1-week's use of a rechargeable LED lamp. At the end of the week, they return it and pay the next week's instalment for an already recharged LED light-box replacement. Using a similar approach, villages will also be able to buy communal toilets, with monthly instalments of just 20 Rupees (40 cents) for a period of 5 or 10 years.

M-Pesa and A Little World are part of a new movement of social innovation, led by social entrepreneurs (Miller et al. 2012). Ashoka, a foundation dedicated to nurturing this new breed of change agent, defines social entrepreneurs as those 'who have innovative solutions to social problems and the potential to change patterns across society. They demonstrate unrivalled commitment to bold new ideas and prove that compassion, creativity, and collaboration are tremendous forces for change.' Put more simply, Ashoka founder Bill Drayton (2010) says that 'the life purpose of the true social entrepreneur is to change the world.'

Although social entrepreneurs are vital to bringing creativity to CSR 2.0 (Ruebottom 2013), this overlooks the important role of innovation within large companies—what the SustainAbility and Skoll Foundation (2008) called social 'intrapreneurship' (Halme et al. 2012). Within companies, innovation is often driven by 'champions' for sustainability or social responsibility causes (Visser 2008a). My research has identified four types of sustainability champion, based on how these change agents derive satisfaction from their work: experts, facilitators, catalysts and activists (Visser 2008d). Similarly, Schaubroeck et al. (2012) emphasise that ethical leadership needs to be embedded within and across organization levels.

In other research, I have found that sustainability innovation tends to come from innovation processes specifically targeting sustainability issues, rather than efforts at embedding sustainability principles in core innovation processes (Blowfield et al. 2007). This is a fundamental distinction, because it means that most Research & Development (R&D) going on in companies—and hence most innovation—is not systematically integrating social and environmental criteria. As a result, much like CSR more generally, innovation is a peripheral, project or product specific activity, which is exactly what is preventing scalable solutions from emerging in the mainstream economy. Until CSR is built into every organisational process—and especially into strategic functions like R&D or new product development—we will always be playing on the fringes of the age of responsibility.

5.2 Scalability

There has been a growing trend of 'ethical consumerism' over the past 50 years (White et al. 2012). Let's look at some of the facts. The UK Soil Association launched the world's first organic standard in 1967 and Germany launched its Blue Angel eco-label in 1978. The first fairtrade coffee, introduced by the Max Havelaar Foundation, was in 1988, and the Rainforest Alliance launched its Smart Wood certification in 1989. Today, in the UK, where the proportion of ethical consumers is among the highest in the world, a survey of 4,000 consumers by PwC found that shoppers buying Fairtrade products rose from 20 % in 2005 to 50 % in 2008 and organic food purchasing increased from 22 % to 43 % over the same period.

However, this £300 billion sector accounted for just 4 % of the UK retail market in 2008 and only 60 % of basic grocery products had sustainable alternatives, falling to 40 % for some sub categories, such as clothing and non-food items. According to the PwC (2008) survey, the high prices associated with fairtrade and organic products remain the main inhibitor to further growth. On average, the price premium for environmentally and ethically-friendly products—taken across 75 items at the UK's top six grocers—was 45 %. Almost 50 % of those shoppers surveyed said they were unwilling or unable to pay this premium, claiming that on average they were not willing to pay a premium in excess of 20 % for greener alternatives.

Simply put, by creating a premium-priced, niche market for 'ethical consumption' (whether it be organic, fairtrade or eco-friendly), companies have been able to present a responsible front to the world, while leaving the vast majority of their products—which are, by implication, less than ethical—unquestioned and unchanged (Bezawada and Pauwels 2013). At the same time, a small group of usually well-off Western consumers have been able to ease their conscience by feeling that they are making a positive difference (Peloza et al. 2013). As a result, these products have never gone to scale. As compared with the total and on-going impacts of mainstream shopping habits, ethical consumption, laudable as it is, has remained marginal at best and totally insignificant at worst (Doherty et al. 2013).

So how do we achieve scalability? One way is by applying so-called 'choice editing' in the supply chain. In fact, manufacturers and retailers choice edit all the time—for example on quality, price, aesthetics and brand. The only difference is that now we are asking them to add sustainability and responsibility to their list of criteria. A classic example is Walmart, which has helped achieve scalability in three areas: sustainable fish, cotton and energy-efficient lightbulbs (Visser 2011). Here, we look briefly at the first two categories.

In 2006, Walmart committed to purchase *all* of its wild-caught fresh and frozen fish for the U.S. market from Marine Stewardship Council (MSC) certified fisheries. They are also working work with Global Aquaculture Alliance (GAA) and Aquaculture Certification Council (ACC) to certify that all foreign shrimp suppliers adhere to Best Aquaculture Practices standards in the U.S. Speaking to the *Wall Street Journal*, George Chamberlain, president of the Aquaculture Alliance put the move in perspective: 'The endorsement drew attention; Walmart buys more shrimp than any other U.S. company, importing 20,000 tons annually—about 3.4 % of U.S. shrimp imports. With Walmart's nod, we went from trying to convince individual facilities to become certified to having long waiting lines.'

Scott also made a commitment to phase out chemically-treated textile crops. By 2008, Wal-Mart was the largest buyer of organic cotton, with more than 10 million pounds purchased annually. They are also the world's largest purchaser of conversion cotton—cotton grown without chemicals, but waiting to be certified as organic. Scott is under no illusions about the ripple effects: 'Cotton farmers can now invest in organic farming because they have the certainty and stability of a major buyer. Through leadership and purchasing power, all of us can create new markets for sustainable products and services. We can drive innovation. We can build acceptance. All we need is the will to step out and make the difference.'

Another way to achieve scalability is through government policy. Consider the way in which governments finally intervened in most countries in order to ban smoking in public places. What does this teach us? First, we know that it took decades of scientific research to make the health case irrefutable. Second, the reforms were helped along by a major exposé of the lies and manipulation by big tobacco companies to undermine progressive legislation. Third, it took the weight of a major public body like the United Nations—in this case the World Health Organization—to legitimise the findings of the scientists. And lastly, it took courage by the politicians to take strong action that was so clearly in the public's best interest, but would inevitably attract big-bully lobbying from the tobacco, restaurant and pub industries.

The interesting thing is that we have all but the fourth ingredient already in place for the broader sustainability and responsibility agenda: the science and research is mounting and calling for urgent action; numerous companies have been exposed for wanton self-aggrandisation; and the UN and other major bodies are putting their weight behind reforms. All that is still lacking is political courage, and even there, we see some signs of movement. For example, the UK's climate change targets, enshrined in the Climate Change Act, are nothing if not ambitious, committing the country to reduce its carbon emissions by 80 % by 2050, with an interim target of 34 % cuts by 2030.

The simple fact is that—beyond a few products like iPods that create their own mass market driven first by innovation and then by clever marketing and latent customer demand—most sustainable and responsible products and services need bold government intervention to make them competitive and scalable. There are at least two good reasons for this. First, as we have already seen, many of the less sustainable and responsible markets are being subsidised, so there is no level playing field on which to compete. And second, most of the more sustainable and responsible products and services that need scaling stray into the territory of being public goods, where the market fails. Speaking to me in 2008, Amory Lovins claimed that there are 'sixty to eighty well-known market failures to buying energy and resources efficiency'. Put another way, if we want CSR to be scalable, smart government regulation is absolutely essential.

5.3 Responsiveness

The problem with CSR 1.0 approaches is that—despite often being enough to keep regulators and NGOs at bay—they are not adequate for reversing the most serious social and environmental trends. For example, a company may show that it has a waste management programme, or a climate change strategy, while total waste and carbon emissions continue to rise (Pinkse and Kolk 2012). Hence, although they are responding on CSR issues, they are not genuinely *responsive* to the scale and urgency of stakeholder needs (Crilly et al. 2012).

An example of what responsiveness means is the way in which Anglo American tackled the HIV/AIDS crisis in South Africa. In South Africa, where someone dies of AIDS every 2 minutes and almost one-in-three women aged 25–29—and more than a quarter of men aged 30–34—are living with the HIV virus, on average 17 % of employees are infected with the virus. Anglo American has been at the forefront of the war against the disease for nearly three decades (Carroll 2012). Their Group Medical Consultant, Dr Brian Brink, was there right at the beginning—in 1980 he was set the task of discovering the first black South African that had contracted the disease—and he is still battling the scourge.

Their great leap of responsiveness came back in 2002 when Anglo American decided to go beyond simple AIDS awareness programmes and to offer their employees free access to the life saving antiretroviral treatments (ARTs) that had become available. In a 2010 interview with the UK's *Telegraph*, Brink recalls: 'We decided to make the treatment available to all of our staff, despite the fact we didn't know what this would cost. Doing this was transformational and it solved a significant problem for the company—the fact that a lot of our staff were dying.'

Today, Anglo American have a much better handle on both the infection rates and the costs. The company estimates that approximately 12,000 of its 71,000 workforce are currently HIV positive. That is still a chronic situation, but compared with where they started—essentially training up two men for each job in the hope that one of them will survive—they have come a long way. Anglo American spends 3.4 % of its payroll of the HIV/AIDS programme, a figure that will probably go up as HIV positive employees survive longer. The only way to reverse the upward trend is by stopping the new HIV infections.

When Anglo American first committed to offer the ARTs, it was a time when no other company in South Africa was doing so and when the 'business case' had yet to be quantified. Nevertheless, Brink made the persuasive argument to top management that 'purchasing anti-retroviral drugs isn't a cost that's going to kill the company, it's a cost that's going to protect the company'. They were convinced—as much by the moral case as the intuitively sound economic rationale—and today the evidence proves that Brink was right.

At Anglo American, the fully-accounted for cost of treatment is $126 per HIV positive employee. However, people on ARTs are more able to work. Therefore, absenteeism declines 1.9 days per employee per month, which saves of $96 a month. The use of healthcare services also declines, saving $87. Added to this is the fact that staff turnover and benefits payments are reduced, which saves a further $36 a month. At the individual level, the total savings of $219 per patient per month amount to approximately 174 % of the cost of providing treatment.

Another example of impressive responsiveness is Unilever's Sustainable Living Plan, launched in November 2010 by CEO Paul Polman and commiting the company to double the size of the business while reducing their environmental footprint and increasing the positive contribution which they make to society (Ignatius 2012). Polman concedes that 'For those things which we find hard to put a price on—biodiversity, carbon, natural capital—the market has failed us. As a result we live in a world where temperatures are rising, natural resources are being depleted, species loss is accelerating and the gap between rich and poor is increasing. This is completely unsustainable.'

Polman believes that business has to decide what role it wants to play. Does it sit on the sidelines waiting for governments to take action or does it get on the pitch and start addressing these issues? 'In Unilever', he says, 'we believe that business must be part of the solution. But to be so, business will have to change. It will have to get off the treadmill of quarterly reporting and operate for the long term. It will have to see itself as part of society, not separate from it. And it will have to recognise that the needs of citizens and communities carry the same weight as the demands of shareholders.'

What distinguishes the plan is the scale of its ambition. For example, by 2020, Unilever aims to: help more than a billion people to improve their hygiene habits; bring safe drinking water to 500 million people; halve the greenhouse gas impact of their products across the lifecycle; halve the water associated with the consumer use of their products; halve the waste associated with the disposal of their products; source 100 % of their agricultural raw materials sustainably; and engage with at least 500,000 smallholder farmers and 75,000 small-scale distributors in their supply network.

Importantly, Unilever believes that responsiveness is linked to long term prosperity. Polman concludes that 'in future this will become the only acceptable model of business. If people feel that the system is unjust and does not work for them, they will rebel against it. And if we continue to consume key inputs like water, food, land and energy without thought as to their long-term sustainability, then none of us will prosper.'

5.4 Glocality

The term 'glocal'—a portmanteau of global and local—is said to come from the Japanese word 'dochakuka', which simply means global localisation. Originally referring to a way of adapting farming techniques to local conditions, dochakuka evolved into a marketing strategy when Japanese businessmen adopted it in the 1980s. It is said that the English word 'glocal' was first coined by Akio Morita, founder of Sony Corporation. In fact, in 2008, Sony Music Corporation even trademarked the phrase 'go glocal'. It was subsequently introduced and popularised in the West in the 1990s by Manfred Lange, Roland Robertson, Keith Hampton, Barry Wellman and Zygmunt Bauman.

In a CSR context, the idea of 'think global, act local' recognises that most CSR issues manifest as dilemmas, rather than easy choices. In a complex, interconnected CSR 2.0 world, companies (and their critics) will have to become far more sophisticated in understanding local contexts and finding the appropriate local solutions they demand, without forsaking universal principles. It is also a caution against applying global models and standards, without allowing for the flexibility of local adaptation and expression. This principle celebrates the diverse ways that CSR is manifesting in different regions and countries of the world.

My work on CSR glocality began when I examined Archie Carroll's (1991) classic CSR pyramid of economic, legal, ethical and philanthropic responsibilities to see how well it fitted the African context (Visser 2006). My conclusion was that 'the relative priorities of CSR in Africa are likely to be different from the classic, American ordering' and that 'Carroll's CSR pyramid may not be the best model for understanding CSR in general, and CSR in Africa in particular.' I then broadened this to propose an alternative CSR pyramid for developing countries, which I outline briefly below and in Fig. 5.1 (Visser 2008b).

Of course, I was not the first to question Carroll's model. Leading CSR academics, Crane and Matten (2010) observed that 'all levels of CSR [described in Carroll's pyramid] play a role in Europe, but they have different significance, and furthermore are interlinked in a somewhat different manner'. In the same way, my contention is that the order of the CSR layers in developing countries—if this are taken as an indicator of the relative emphasis assigned to various responsibilities—differs from Carroll's classic pyramid. Hence, in developing countries, economic responsibilities still get the most emphasis. However, philanthropy is given second highest priority, followed by legal and then ethical responsibilities.

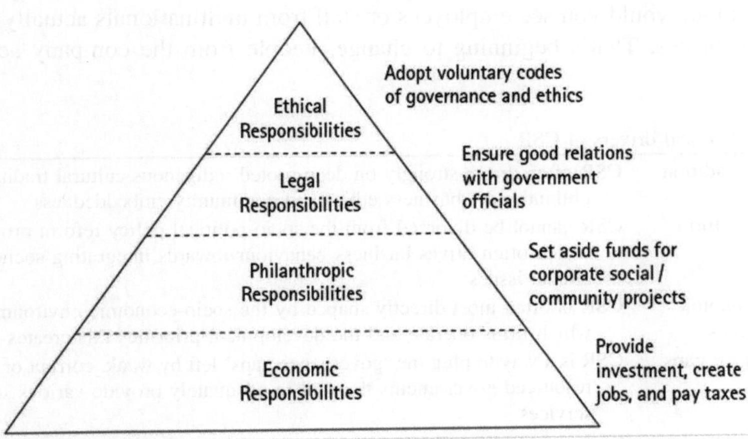

Fig. 5.1 CSR pyramid for developing countries

Beyond my research of a CSR Pyramid for developing countries, the importance of glocality for CSR really struck home to me when, in 2008—together with my co-author, Dirk Matten—I launched my book *The A to Z of Corporate Social Responsibility* in several regions and countries around the world, from Guatemala and South Africa to China and the UK (Visser et al. 2007). What became blindingly obvious was that while CSR had some global principles that most countries agreed on, the local manifestations were distinctive in each local case.

This led me to undertake a 2-year research project that culminated in *The World Guide to CSR*, published in 2010 and profiling CSR in five regions and 58 countries (Visser and Tolhurst 2010). In her review of the book, Cohen (2010) captures some of the essence of the idea of glocality: 'The country profiles [offer] a local flavour and sometimes even a little local language—*tzedakah*, the Hebrew word for charity; *sanpo yoshi*—'three-way good' in Japan; *choregia*, the ancient form of sponsorship in Greece; and *ubuntu* in Southern Africa, which relates to community culture, to name but a few examples.'

One of the ways glocality is determined is that each region, country or community has a different combination of CSR drivers. The art of glocality, therefore, is to determine which incentives and pressures are most applicable to the local context. Of the 10 typical drivers I have identified, five are local (or internal) drivers, namely pressures from within the country or community. These are briefly discussed in Tables 5.2 and 5.3.

Other sets of drivers are more global (or external) and tend to have an international origin. Remember, it is the varied combination of drivers that determines glocality.

Making these glocal drivers work in practice requires a special skill, which Stuart Hart (2007) calls 'native capability'. When I interviewed him (Visser and CPSL 2009b), he explained it like this: 'Where multinationals have engaged with Base of the Pyramid (BOP) communities, in rural areas or slums and shanty towns, it's almost always been through an NGO partner; they outsourced it to that partner. Very seldom would you see employees or staff from multinationals actually being in those places. That's beginning to change. People from the company actually

Table 5.2 Local drivers of CSR

Cultural tradition	CSR often draws strongly on deep-rooted indigenous cultural traditions of philanthropy, business ethics and community embeddedness
Political reform	CSR cannot be divorced from the socio-political policy reform process, which often drives business behaviour towards integrating social and ethical issues
Socio-economic priorities	CSR is often most directly shaped by the socio-economic environment in which firms operate and the development priorities this creates
Governance gaps	CSR is a way to plug the 'governance gaps' left by weak, corrupt or under-resourced governments that fail to adequately provide various social services
Crisis response	CSR responses can be catalysed by economic, social, environmental, health-related or industrial crises

Table 5.3 Global drivers of CSR

Market access	CSR may be seen as an enabler for companies in one country or region trying to access markets in other parts of the world
International standardisation	CSR codes, guidelines and standards are a key driver for companies wishing to operate as global players
Investment incentives	CSR is given an incentive by the trend of socially responsible investment (SRI), where funds are screened on ethical, social and environmental criteria
Stakeholder activism	CSR is encouraged through the activism of stakeholders or pressure groups, acting to address the perceived failure of the market and government policy
Supply chain integrity	CSR activities among small and medium-sized companies are boosted by requirements imposed by multinationals on their supply chains

have to be in those spaces and develop relationships on the ground. You can't just outsource the work to NGOs and expect any capability to develop (Hart 2008).

'So it's really a new capability,' says Hart, 'the idea of native capability or becoming indigenous, and it requires effort to develop it. But if you do it and do it well, it can yield new business opportunities based on trust and social capital that makes you virtually impossible to dislodge. So from the standpoint of sustainable competitive advantage, becoming embedded in the community, developing that sort of social capital and trust and relationship, is the highest form of sustainable competitive advantage' (Hart 2008; Ansari et al. 2012).

5.5 Circularity

The need for a circular economy has origins in the 1960s, when the popular metaphor of 'Spaceship Earth' emerged, led by economists Barbara Ward (1966) and Kenneth Boulding (1966) and polymath R. Buckmister Fuller (1969). For instance, Boulding contrasted the 'cowboy economy', an open economy with seemingly boundless frontiers and limitless resources, with the 'spaceman economy', a closed economy of the future 'in which the earth has become a single spaceship, without unlimited reservoirs of anything, either for extraction or for pollution, and in which, therefore, man must find his place in a cyclical ecological system.'

At a more pragmatic level, there was also a movement since the 1960s that led to the modern practice of Life Cycle Assessment (LCA). As far back as 1969, a study by Coca-Cola assessed different beverage containers to determine which had the lowest releases to the environment and least affected the supply of natural resources. This emergent methodology became known as a Resource and Environmental Profile Analysis (REPA) in the U.S. and as an Ecobalance in Europe. The result of all of these developments is that today LCA is widely understood and employed—often using software like SIMAPRO—as 'the assessment of the environmental impact of a product in a cradle-to-grave approach', whereby 'all the pollution from the stage of digging or harvesting raw materials to the waste that remains after using a product is taken into account' (De Bruijn 2010).

Towards the end of the 1980s, a related concept called 'industrial ecology' emerged. It was popularized in a *Scientific American* article by Frosch and Gallopoulos (1989), in which they declared: 'Why would not our industrial system behave like an ecosystem, where the wastes of a species may be resources to another species? Why would not the outputs of an industry be the inputs of another, thus reducing use of raw materials, pollution, and saving on waste treatment?' An example is the Danish industrial park in the city of Kalundborg where a power plant, oil refinery, pharmaceutical plant, plasterboard factory, enzyme manufacturer, waste management company and the city itself all link together to share and utilise resources, by-products, energy and waste heat.

Another concept that was gaining popularity around the same time is 'cleaner production', which resulted in the UNEP Declaration on Cleaner Production in 1998. Later, this evolved into the concept of 'sustainable consumption and production', which was defined at the UN's 2002 World Summit on Sustainable Development as an approach 'to promote social and economic development within the carrying capacity of ecosystems by addressing and, where appropriate, delinking economic growth and environmental degradation through improving efficiency and sustainability in the use of resources and production processes and reducing resource degradation, pollution and waste.'

One of the most integrated and powerful methodologies to emerge from all of these trends towards circularity is 'cradle to cradle'. While the phrase was coined by Walter R. Stahel in the 1970s, its modern interpretation grew out of a system of 'lifecycle development' initiated by Michael Braungart and colleagues at the Environmental Protection Encouragement Agency (EPEA) in the 1990s and explored through the publication 'A Technical Framework for Life-Cycle Assessment'. Braungart (2008) told me how it all got started: 'I was looking at complex household products and I identified in the TV set 4,360 different chemicals, and I thought it doesn't help just to take any toxic stuff out of it.'

The concept was popularised when Braungart teamed up with U.S. architect William McDonough and produced the book *Cradle to Cradle: Remaking the Way We Make Things* (McDonough and Braungart 2009). The authors explain that cradle to cradle goes beyond 'cradle to grave' thinking. Hence, rather than simply considering impacts across the life cycle of a product and trying to minimise waste, the authors argue for closed-loop production, where waste is only acceptable if it is entirely re-used by the system. Hence, all waste becomes 'food' input to the cycles of nature and the cycles of industry, i.e. either as a biological nutrient or a technical nutrient. McDonough elaborates: 'The idea is that you would design something as a biological nutrient, so that it can go back to the soil to refresh the soil and rebuild the soil, rather than be wasted or toxify the soil through the air and the water. And then other things we see as technical nutrients that are designed to be in infinite closed cycles in technology and manufacturing.'

Towards the end of the 1990s, the Nike Environmental Action Team (NEAT), led by Sarah Severn, then Nike's Director of Corporate Sustainable Development, began shaping what would become Nike's cradle to cradle approach. Severn liked the innovative and ambitious approach of cradle to cradle: 'So much of the environmental debate had addressed end-of-pipe problems and end-of-pipe solutions,' she recalled. 'And here was a strategy that was turning that on its head. It was not about restriction or reaction. It created positive solutions at the front of the design process' (Visser 2011).

In order to measure progress of all of its products and designs against its cradle to cradle principles, Nike developed a Considered Index, which is a tool for evaluating the predicted environmental footprint of a product prior to commercialisation. This system examines solvent use, waste, materials and innovation for footwear. Apparel products are evaluated on waste, materials, garment treatments and innovation. Nike claim that 'by continually raising that standard, we envision a future where the shoes you wear today become the shoes, shirts or equipment you use tomorrow. This closed loop manufacturing process, where nothing is wasted and everything is kept in play, is not just wishful thinking, it's the future.'

Of course, it is not only textile companies that are embracing the circular economy. In 2007, Coca Cola CEO Neville Isdell pledged that his company would become 'water neutral' (Liu 2008). Fuji Xerox has made similar commitments to zero waste and by 2010, Fuji Xerox Eco-Manufacturing in Asia was able to announce that it had effectively accomplished the Zero Landfill goal by recycling 99.8 % of used products and consumables.

These and many other initiatives are captured in Elkington's (2012) *The Zeronauts*—'a new breed of innovator, determined to drive problems such as carbon, waste, toxics and poverty to zero'. As I concluded in my review of the book (Visser 2012), 'I'm not sure any of the new jargon will stick, and I'm not sure that it matters. The substance of the book—which is about the potential for disruptive innovation in the new 'Race to Zero'—is guaranteed to inspire a new generation of sustainability thinking and practice. And as these new pioneers increasingly perform—to quote *Alice in Wonderland*—'as many as six impossible things before breakfast', we may find ourselves looking back in decades to come and wondering why the Sustainability Barrier seemed, for so many years, impossible to break.'

References

Ansari S, Munir K, Gregg T (2012) Impact at the 'bottom of the pyramid': the role of social capital in capability development and community empowerment. J Manage Stud 49(4):813–842

Arora B, Metz Cummings A (2010) A little world: safe and efficient m-banking in rural India. UNDP Growing Inclusive Markets publication, New York

Bezawada R, Pauwels K (2013) What is special about marketing organic products? how organic assortment, price, and promotions drive retailer performance. J Mark 77(1):31–51

Blowfield M, Visser W, Livesey F (2007) Sustainability innovation: mapping the territory, University Cambridge Programme for Sustainability Leadership Research Paper Series, No. 2

Boulding KE (1966) The economics of the coming spaceship earth. In: Jarrett H. (ed) Environmental quality in a growing economy, Resources for the Future/Johns Hopkins University Press, Baltimore, pp 3–14

Braungart M (2008) Interview with Michael Braungart, conducted by Wayne Visser as part of the University of Cambridge Top 50 Sustainability Books project

Braungart M, McDonough W (2009) Cradle to cradle: remaking the way we making things. Random House, New York

Carroll AB (1991) The pyramid of corporate social responsibility: toward the moral management of organizational stakeholders. Bus Horiz 34:39–48

Carroll C (2012) The CEO of anglo American on getting serious about safety. Harvard Bus Rev 90(6):43–46

Cohen E (2010) The world guide to CSR (Book Review). CSRwire, 13 Aug

Crane A, Matten D (2010) Business ethics: managing corporate citizenship and. sustainability in the age of globalization, 3rd edn. Oxford University Press, Oxford

Crilly D, Zollo M, Hansen MT (2012) Faking it or muddling through? understanding decoupling in response to stakeholder pressures. Acad Manag J 55(6):1429–1448

De Bruijn T (2010) Life cycle assessment. In: Visser, Matten, Pohl, Tolhurst (eds) The A to Z of corporate social responsibility. Wiley, London

Doherty B, Davies IA, Tranchell S (2013) Where now for fair trade? Bus Hist 55(2):161–189

Drayton B (2010) Tipping the world: the power of collaborative entrepreneurship. What Matters Voices on Society, McKinsey & Company, Chennai

Elkington J (2012) The zeronauts: breaking the sustainability barrier. Routledge, New York

Esty DC, Charnovitz S (2012) Green rules to drive innovation. Harvard Bus Rev 90(3):120–123

Frosch RA, Gallopoulos NE (1989) Strategies for manufacturing. Sci Am 189(3):144–152

Fuller RB (1969) Operating manual for spaceship earth. Simon & Schuster, New York

Halme M, Lindeman S, Linna P (2012) Innovation for inclusive business: intrapreneurial bricolage in multinational corporations. J Manage Stud 49(4):743–784

Hart, SL (1997) Beyond greening: strategies for a sustainable world. Harvard Bus Rev 75(1):66–76

Hart SL (2007) Capitalism at the crossroads: aligning business, earth and humanity. Wharton School Publishing, New Jersey

Hart (2008) Interview with Stuart Hart, conducted by Wayne Visser as part of the University of Cambridge Top 50 Sustainability Books project, 13 Aug 2008

Hollender J, Breen B (2010) The responsibility revolution: how the next generation of businesses will win. Jossey-Bass, San Francisco

Ignatius Adi (2012) Captain Planet. Harvard Bus Rev 90(6):112–118

Liu LW (2008) Water pressure. Time Magazine, 12 June

McKinsey (2010). How companies manage sustainability. Research report

Miller TL, Grimes MG, McMullen JS, Vogus TJ (2012) Venturing for others with heart and head: how compassion encourages social entrepreneurship. Acad Manag Rev 37:616–640

Nidumolu R, Prahalad CK, Rangaswami MR (2009) Why sustainability is now the key driver of innovation. Harvard Bus Rev 87(9):56–64

Peloza J, White K, Jingzhi S (2013) Good and guilt-free: the role of self-accountability in influencing preferences for products with ethical attributes. J Mark 77(1):104–119

Pinkse J, Kolk A (2012) Multinational enterprises and climate change: exploring institutional failures and embeddedness. J Int Bus Stud 43(3):332–341

PwC (2008) Sustainability: are consumers buying it? PricewaterhouseCoopers report

Ruebottom T (2013) The microstructures of rhetorical strategy in social entrepreneurship: building legitimacy through heroes and villains. J Bus Ventur 28(1):98–116

Schaubroeck JM, Hannah ST, Avolio BJ, Kozlowski SW, Lord RG, Trevinño LK, Dimotakis N, Peng AC (2012) Embedding ethical leadership within and across organization levels. Acad Manag J 55(5):1053–1078

Sengupta A (2012) Investment in cleaner technology and signaling distortions in a market with green consumers. J Environ Econo Manag 64(3):468–480
SustainAbility and Skoll Foundation (2008) The social intrapreneur: a field guide for corporate changemakers. SustainAbility, London
Visser W (2006) Revisiting carroll's CSR pyramid: an african perspective. In: Pedersen ER, Huniche M (eds) Corporate citizenship in developing countries. Copenhagen Business School Press, Copenhagen, pp 29–56
Visser W (2008a) Change agents—happiness is about making things better. Ethical Corporation, 12 Nov
Visser W (2008b) Making a difference: purpose-inspired leadership for corporate sustainability & responsibility. VDM, Saarbrücken
Visser W (2008c) Corporate social responsibility in developing countries. In: Crane A, McWilliams A, Matten D, Moon J, Siegel D (eds) The Oxford handbook of corporate social responsibility. Oxford University Press, Oxford, pp 473–479
Visser W (2009) CPSL (Cambridge Programme for Sustainability Leadership) 2009b. The top 50 sustainability books, Greenleaf Publishing, Sheffield
Visser W, (2011) The age of responsibility: csr 2.0 and the new DNA of business. John Wiley & Sons, London
Visser W (2012) The quest for sustainable business: an epic journey in search of corporate responsibility. Greenleaf, Sheffield
Visser W (2013) Corporate sustainability & responsibility: an introductory text on csr theory & practice—past, present & future. Kaleidoscope Futures, London
Visser W, Tolhurst N (2010) The world guide to CSR. Greenleaf, Sheffield
Visser W, Matten D, Pohl M, Tolhurst N (eds) (2007) The a to z of corporate social responsibility: a complete reference guide to concepts, codes and organisations. Wiley, London
Ward B (1966) Spaceship earth. Columbia University Press, USA
White K, MacDonnell R, Ellard JH (2012) Belief in a just world: consumer intentions and behaviors toward ethical products. J Mark 76(1):103–118
Xueming L, Shuili D (2012) Good companies launch more new products. Harvard Bus Rev 90(4):28

Chapter 6
Assessing CSR 2.0 Performance

Abstract A 70-question self-assessment diagnostic tool is presented to quantify the five stages of CSR, the five principles of CSR 2.0 and the four DNA elements of CSR 2.0. Sample data is used to show how the tool can be employed for future research and practitioner application, including an illustrative report to management that describes areas for future improvement.

Keywords Corporate social responsibility · CSR 2.0 · Tool · Assessment · Diagnostic · Measurement · Indicators · Reporting · Benchmarking · Data

Based on the frameworks presented in this book, in 2011 I created, in collaboration with technical partners Hexagon Consultores, a 70-question online CSR 2.0 Self Assessment Diagnostic Tool. The Diagnostic evaluates organisations against:

- The five stages of CSR (defensive, charitable, promotional, strategic and transformative)
- Five principles of CSR 2.0 (creativity, scalability, responsiveness, glocality, circularity)
- The four DNA elements of CSR 2.0 (value creation, good governance, societal contribution and environmental integrity)

Although the diagnostic is primarily designed as a tool for practitioners—in fact, consultants can be trained and licenced as "CSR 2.0 Assessors and Advisors" using the diagnostic—it nevertheless offers significant potential as a questionnaire for applied research, and academics are invited to contact me if they which to use it as part of a research project.

In this section, I will present real data from a workshop in which the diagnostic was administered for 'Organisation X'. The data is captured using a Likert-type scale and converted into percentage scores. Note, however, that *the data presented below is not statistically significant and is used purely for illustrative purposes.* Also be aware that, in practice, the questions appear randomly to reduce respondent bias.

W. Visser, *CSR 2.0*, SpringerBriefs in Organisational Studies,
DOI: 10.1007/978-3-642-40874-8_6, © The Author(s) 2014

6.1 Stages of CSR

The CSR 2.0 diagnostic measures the evolution of business responsibility in terms of five different stages of CSR, namely: defensive, charitable, promotional, strategic and transformative CSR. These stages have been described in more detail in Chap. 2.

Companies tend to move through these stages of maturity, although they may have activities in several stages at once. The aspiration is to progressively improve performance in each subsequent stage, ultimately practicing transformative CSR, which is also called CSR 2.0. By way of a reminder, the stages are summarised in Table 6.1. The self-assessment scores of Organisation X for these stages are shown in Fig. 6.1.

Table 6.1 The stages of CSR

Dominant paradigm	Stage of CSR	Modus operandi	Key enabler	Stakeholder target
Greed	Defensive	Ad hoc interventions	Investments	Shareholders, government & employees
Philanthropy	Charitable	Charitable programmes	Projects	Communities
Marketing	Promotional	Public relations	Media	General public
Management	Strategic	Management systems	Codes	Shareholders & NGOs/CSOs
Responsibility	Transformative	Business models	Products	Regulators & customers

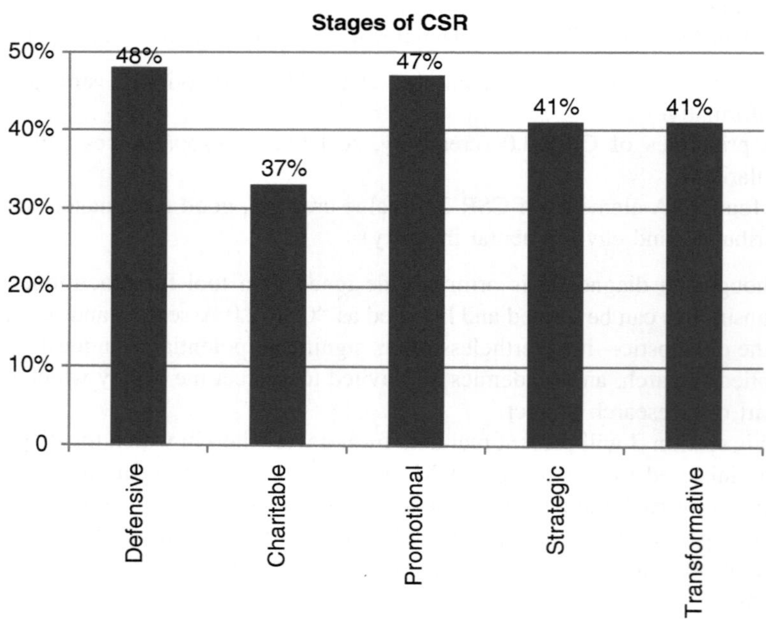

Fig. 6.1 Stages of CSR (illustrative scores)

6.1.1 Defensive CSR

Figure 6.1 shows that Organisation X performs most strongly on defensive CSR. This typically reflects a corporate culture that is very focused on economic value. This is confirmed by the fact that most respondents agree 'completely' or 'to a large extent' that the organisation's growth, profitability and/or shareholder returns are the key measure of success, and that staff performance appraisals are linked to the economic performance of their unit or of the organisation. CSR practices during this stage are usually seen as a risk management tool to minimise stakeholder complaints, government regulation and fines or penalties for legal non-compliance (Table 6.2).

Making the 'business case' for CSR (i.e. justifying it in terms of financial benefits) is very important for organisations in this stage of CSR. The danger is that CSR practices remain narrow and reactive. Practical steps for improving in this stage include making a public commitment to long term shareholder value (rather than short term quarterly returns) and engaging with socially responsible investment (SRI) indexes and funds (Hong and Kostovetsky 2012).

6.1.2 Charitable CSR

Organisation X performs most weakly on charitable CSR, although this may simply reflect that philanthropic donations are not institutionalised through a Foundation, Trust or Chairman's Fund. Despite the lack of formal structures, most

Table 6.2 Defensive CSR (illustrative scores)

Diagnostic questions	Score (%)
To what extent does your organisation believe that CSR should be entirely voluntary, as an alternative to greater government regulation	41
To what extent is your organisation's growth, profitability and/or return to shareholders the key measure of success	64
To what extent are your staff performance appraisals linked to the economic performance of their unit or of the organisation	57
To what extent is your organisation oriented towards short term (quarterly) financial results, rather than long term societal goals	43
To what extent is your organisation's culture perceived to be about tactical cost cutting, rather than strategic value creation	35
Average	48

Table 6.3 Charitable CSR (illustrative scores)

Diagnostic questions	Score (%)
To what extent does your organisation look to the generous spirit of its owner/CEO/chairman to take a lead on charity	45
To what extent is your organisation's culture built on the idea of making a contribution to the community and giving back to society	60
To what extent are employees given paid volunteer days (in which they can be directly involved in charitable projects)	25
To what extent is charitable giving institutionalised, e.g. through a foundation or Chairman's fund	19
To what extent is the organisation's CSR activities managed by staff that are primarily responsible for charitable donations or community projects	35
Average	37

respondents agree 'completely' or 'to a large extent' that the organisation's culture is built on the idea of making a contribution to the community and giving back to society (Table 6.3).

Practical steps for improving in this stage include giving employees paid volunteer days, in which they can be directly involved in charitable projects (Grant 2012), employing impact investment techniques and engaging in strategic philanthropy (aligning causes to the core business) and venture philanthropy (investing in social enterprises).

6.1.3 Promotional CSR

Promotional CSR is rated second most highly for Organisation X. Most respondents agree 'completely' or 'to a large extent' that CSR is used primarily as a means to enhance the organisation's brand equity, public reputation or stakeholder relations. Despite this, most believe that the organisation ranks on CSR ratings and awards (e.g. CSR reporting awards, CSR indexes) only 'to a small extent' or 'not at all'. Despite obvious public relations benefits, the danger with the marketing approach to CSR is that the organisation may be accused of 'spin', or 'greenwash', i.e. making responsibility or sustainability claims that it cannot justify in terms of the actual overall performance of the organisation (Kronrod et al. 2012) (Table 6.4).

Table 6.4 Promotional CSR (illustrative scores)

Diagnostic questions	Score (%)
To what extent does your organisation manage to obtain media coverage for its CSR activities	55
To what extent is your organisation's public relations, corporate affairs or marketing departments responsible for CSR efforts	53
To what extent is CSR seen as a means to enhance the organisation's brand equity, public reputation or stakeholder relations	69
To what extent does your organisation rank on CSR ratings and awards (e.g. CSR reporting awards, CSR indexes)	19
To what extent does your CEO talk publicly or to the media about the organisation's CSR performance	37
Average	47

Practical steps for improving in this stage include commitment to GRI reporting, life cycle product-based reporting and eco-labelling, as well as using cause-related or social marketing techniques to simultaneously improve transparency, social impact and sales (Koschate-Fischer et al. 2012; Robinson et al. 2012).

6.1.4 Strategic CSR

Organisation X performs equally on strategic CSR and transformative CSR. In terms of Strategic CSR, most respondents agree 'completely' or 'to a large extent' that the CSR issues that the organisation supports are aligned to its core business. However, it seems the organisation is not certified against internationally recognised CSR standards like ISO 14001, OHSAS 18001 and SA 8000, nor is its CSR performance audited by independent third parties (e.g. auditors, consultants, certifiers).

Practical steps for improving in this stage include engaging more formally (through codes and standards) with environmental management, supply chain management and strategic issue management, which typically involve cycles of CSR policy development, goal and target setting, programme implementation, auditing and reporting (Table 6.5).

Table 6.5 Strategic CSR (illustrative scores)

Diagnostic questions	Score (%)
To what extent are the CSR issues that your organisation supports aligned to its core business	69
To what extent is CSR embedded through internal management systems (policies, objectives, targets, procedures, reviews & reports)	49
To what extent can your organisation demonstrate quantified continuous improvement on social, environmental & ethical performance	48
To what extent is your organisation's CSR performance audited by independent third parties (e.g. auditors, consultants, certifiers)	25
To what extent is your organisation certified against internationally recognised CSR standards like ISO 14001, OHSAS 18001 and SA 8000	15
Average	41

6.1.5 Transformative CSR (or CSR 2.0)

Although it seems that the mission of Organisation X is at least partially about solving a particular environmental, social or ethical challenge, respondents scored the company fairly low in terms of its performance against CSR goals and its ambition for radical change. Transformative CSR is about identifying and tackling the root causes of our present unsustainability and irresponsibility, typically through innovating business models, revolutionising their processes, products and services and lobbying for progressive national and international policies.

Practical steps for improving in this stage include conducting full life cycle social and environmental impact assessments for the organisation's products and services, and setting bold CSR targets to inspire action (Table 6.6).

Table 6.6 Transformative CSR (illustrative scores)

Diagnostic questions	Score (%)
To what extent does your organisation believe it is falling short of the goals of sustainability and responsibility	39
To what extent is your organisation's mission about solving a particular environmental, social or ethical challenge	51
To what extent is your industry or sector perceived to have minimal social, environmental or ethical costs and/or risks	41
To what extent is your organisation's CSR performance recognised as a key differentiator in the markets that it operates in	45
To what extent has your organisation undergone radical strategic changes over the past 12 months as a result of social, environmental or ethical issues	29
Average	41

6.2 Principles of CSR 2.0

Transformative CSR—which is also called CSR 2.0—is based on five principles: creativity, scalability, responsiveness, glocality and circularity. These hold the key to making change happen, at a societal, organisational and individual level, and ensuring that we can all make a difference. The principles have been described in some detail in Chap. 5. As a reminder, however, Table 6.7 summarises the main features of the five principles (Fig. 6.2).

Table 6.7 Principles of CSR 2.0

Principle	Description	Examples
Creativity	Moving beyond 'tick-box' CSR approaches like codes & standards to embrace social & environmental innovation & entrepreneurship, creating products & services to solve sustainability challenges	Freeplay Energy's 'wind-up' energy technologies (radios, torches, laptops); Vodafone's M-Pesa scheme for money transfer by text
Scalability	Moving beyond 'ethical consumerism' CSR approaches like fairtrade & eco-labelling to embrace 'choice editing' where only sustainable & responsible options are offered for entire product or service lines	Walmart's Marine Stewardship Council (MSC) certified fish product lines; Co-operative Bank's ethical banking & insurance services
Responsiveness	Moving beyond 'shareholder-driven' CSR approaches like charity donations & promotional campaigns to embrace stakeholder responsiveness, including impact investing & stakeholder feedback mechanisms	The Corporate Leaders Group on Climate Change; the Carbon Disclosure Project's climate solutions investment strategy
Glocality	Moving beyond imperial and parochial CSR approaches to embrace 'think global, act local' practices, which follow international guidelines, but tailor context-specific solutions	BHP-Billiton's dual strategies for climate change (global) & malaria prevention (local); Local Agenda 21 actions, based on global principles
Circularity	Moving beyond 'end of pipe' CSR approaches to embrace 'cradle to cradle' practices, which close the loop on production & service processes, with the goal of zero waste, zero toxics & 100 % renewable resources	Fuji-Xerox's product take-back scheme in Asia; Shaw Carpets' closed loop production, end-of-life collection & recycling system

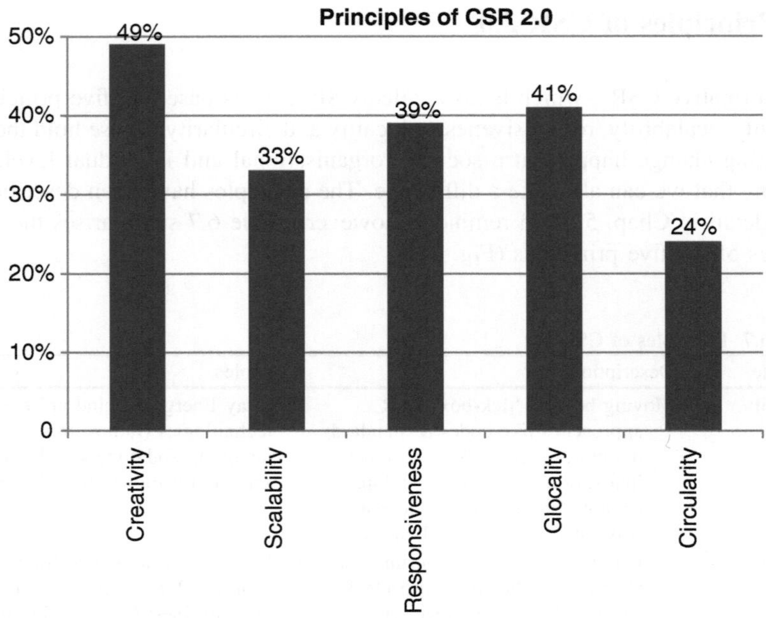

Fig. 6.2 Principles of CSR 2.0 (illustrative scores)

6.2.1 Creativity

Organisation X scores most strongly on fostering a culture of experimentation and opportunity, rather than risk aversion. Practical next steps for improvement could include investing in social enterprises and social entrepreneurs (rather than purely philanthropy), and designing products and services to intentionally solve a social, environmental or ethical problem (Table 6.8).

Table 6.8 Creativity (illustrative scores)

Diagnostic questions	Score (%)
To what extent does your organisation regularly innovate to come up with socially and environmentally responsible solutions or products	51
To what extent are your new products, services and investments deliberately designed to address a social, environmental or ethical challenge	43
To what extent does your organisation support, finance, employ or empower social enterprises or social entrepreneurs	27
To what extent does your organisation encourage a culture of experimentation, rather than penalising mistakes	67
To what extent is CSR seen as a market opportunity or source of revenue, rather than a market risk or source of cost	59
Average	49

Table 6.9 Scalability (illustrative scores)

Diagnostic questions	Score (%)
To what extent does your organisation educate and encourage its customers to shop and live more sustainably and responsibly	41
To what extent are your products and services certified to a recognised social, environmental or ethical label (e.g. fairtrade, organic, etc.)	20
To what extent are your products and services designed to directly serve the bottom of the economic pyramid (i.e. low income markets)	27
To what extent does your organisation engage in choice editing (i.e. only offering social, environmental or ethically screened products)	24
To what extent does your organisation participate in industry association agreements to raise social, environmental or ethical standards in your sector	55
Average	33

6.2.2 Scalability

While Organisation X clearly makes some effort to raise social, environmental or ethical standards through its industry association and by trying to educate its customers, there are still significant areas for improvement. Practical next steps may include pursuing product and service certification according to applicable social, environmental or ethical labelling schemes, starting to implement choice editing for entire product or service categories, and intentionally serving low-income markets (Table 6.9).

6.2.3 Responsiveness

Organisation X performs strongest on cross-sector partnerships and shows some evidence of acting on stakeholder feedback. Practical next steps for improvement are in the area of positive lobbying (for more progressive legal requirements), open source sharing of social and environmental solutions and bringing challenging stakeholder voices onto the board as non-executive directors or advisors (Table 6.10).

Table 6.10 Responsiveness (illustrative scores)

Diagnostic questions	Score (%)
To what extent does your organisation lobby industry bodies and/or government for stronger laws on social, environmental and ethical issues	23
To what extent does your organisation share its social, environmental or ethical solutions through open source forums or patent commons	23
To what extent are non-executive directors with specific expertise in social, environment or ethical issues included on your organisation's board	39
To what extent does your organisation enter into formal cross-sector partnerships (between government, civil society and business)	60
To what extent does your organisation modify its strategy, operations, products/services and projects as a result of stakeholder feedback	52
Average	39

Table 6.11 Glocality (illustrative scores)

Diagnostic questions	Score (%)
To what extent does your organisation give staff members training in the handling of ethical dilemmas	35
To what extent can your organisation demonstrate audited compliance with international social, environmental or ethical codes or standards	25
To what extent do local managers, staff and stakeholders have the ability to determine social, environmental and ethical priorities	47
To what extent does your organisation support local organisations and causes	56
To what extent does your organisation have CSR policies and practices that are adapted and tailored to the local culture and conditions	41
Average	41

6.2.4 Glocality

Organisation X performs strongest on supporting and prioritising local causes. Practical next steps for improvement could include signing up to auditable international codes and standards, as well as training staff in handling ethical dilemmas (Table 6.11).

6.2.5 Circularity

Organisation X appears to have begun addressing the impacts of their products and services, but overall, this principle displays their weakest performance area. Practical next steps might include transitioning to renewable energy supply, implementing product take-back schemes and adopting the goals of zero waste, water neutrality and carbon neutrality (Table 6.12).

Table 6.12 Circularity (illustrative scores)

Diagnostic questions	Score (%)
To what extent is your organisation committed to water neutrality (putting back into nature as much water and of as high quality as it takes out/uses)	13
To what extent is your organisation's energy consumption supplied from renewable sources (wind, solar, nuclear, etc.)	9
To what extent is your organisation achieving the goal of zero waste	33
To what extent does your organisation offer product take back schemes (at the end of their useful life)	15
To what extent does your organisation design products and services to have zero negative social and environmental impacts	49
Average	24

6.3 DNA of CSR 2.0

The CSR 2.0 DNA model—comprising value creation, good governance, societal contribution and environmental integrity—was introduced and described in some detail in Chap. 1. As a reminder, the DNA elements are summarised in Table 6.13. The self-assessment scores against the DNA elements are shown in Fig. 6.3.

Table 6.13 DNA of CSR 2.0

DNA bases	Strategic goals	Example indicators	Description
Value creation	Economic development	Capital investment	Economic, social, human and natural capital
		Beneficial products	Sustainable and responsible goods and services
		Inclusive business	Wealth distribution, bottom of the pyramid markets
Good governance	Institutional effectiveness	Leadership	Strategic commitment to sustainability and responsibility
		Transparency	Sustainability and responsibility reporting, government payments
		Ethical practices	Bribery and corruption prevention, values in business
Societal contribution	Stakeholder orientation	Philanthropy	Charitable donations, provision of public goods andservices
		Fair labour practices	Working conditions, employee rights, health and safety
		Supply chain integrity	SME empowerment, labour and environmental standards
Environmental integrity	Sustainable ecosystems	Ecosystem protection	Biodiversity conservation & ecosystem restoration
		Renewable resources	Tackling climate change, renewable energy and materials
		Zero waste production	Cradle-to-cradle processes, waste elimination

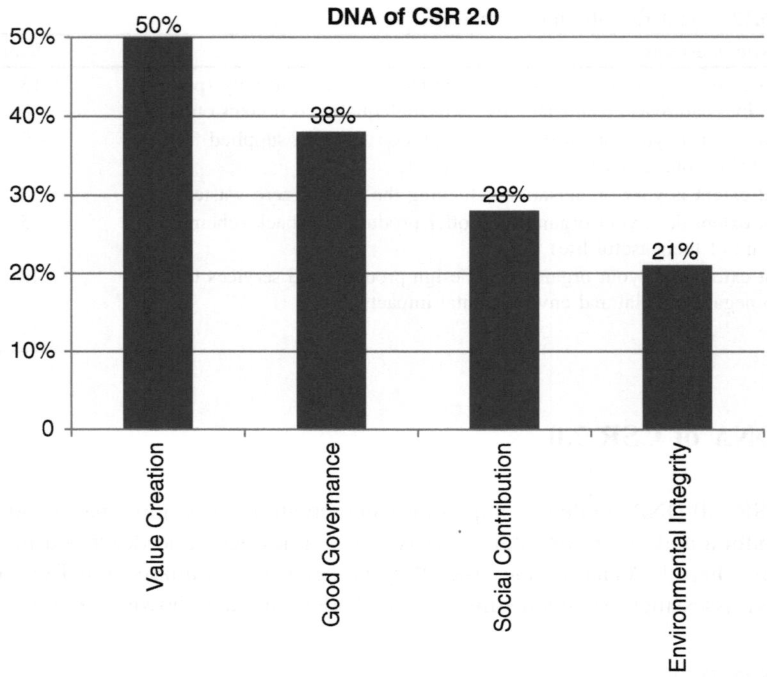

Fig. 6.3 DNA of CSR 2.0 (illustrative scores)

6.3.1 Value Creation

Organisation X performs most strongly on the value creation DNA element of CSR 2.0, although even here, there is considerable scope for improvement. The strongest performance for Organisation X is relatively equitable pay distribution

Table 6.14 Value creation (illustrative scores)

Diagnostic questions	Score (%)
To what extent does your organisation publicly disclose a full and transparent economic value added statement	41
To what extent does your organisation rely on local suppliers, thereby creating indirect jobs (To a Very Great Extent = 100 %)	60
To what extent does is the ratio of pay between CEO/owner and average workers represent an equitable income distribution (To a Very Great Extent = <1:50)	64
To what extent are your products & services socially and environmentally beneficial, with positive effects on health, wellbeing and the environment	55
To what extent are your employees and customers also owners/shareholders of the organisation	31
Average	50

Table 6.15 Good governance (illustrative scores)

Diagnostic questions	Score (%)
To what extent does your organisation publicly commit to and comply with one or more national or international corporate governance codes	41
To what extent does your organisation make full public disclosure of their payments to government (taxes, political party contributions, etc.)	39
To what extent does your organisation produce an externally verified and rated GRI sustainability report (Completely = A + rating)	16
To what extent does your organisation have enforced policies and procedures to prevent and punish bribery, corruption and other unethical behaviour	55
To what extent does your organisation include social, environmental & ethical issues as part of its internal auditing and risk management activities	40
Average	38

and the use of local suppliers, thereby encouraging local economic development. Practical next steps could include considering the introduction of a share ownership scheme, publicly reporting an economic value added statement and looking for ways to tailor products and services to be inclusive of low-income (so-called 'bottom of the pyramid'/BOP) markets (Table 6.14).

6.3.2 Good Governance

Organisation X performs most strongly on having ethics policies, but shows opportunities for improvement in other areas. Practical next steps might include engaging in externally verified reporting based on the Global Reporting Initiative (GRI) standard, publicly committing to follow a national or international corporate governance code, and integrating social, environmental and ethical issues into its internal audit and risk management practices (Table 6.15).

Table 6.16 Societal contribution (illustrative scores)

Diagnostic questions	Score (%)
To what extent does your organisation measure the effectiveness (impact) of its stakeholder dialogues and cross-sector partnerships	28
To what extent are your organisation's suppliers third-party audited on social, environmental & ethical performance	15
To what extent does your organisation contribute a proportion of its pre-tax profits to charitable causes (To a Very Great Extent = >5 %)	29
To what extent are stakeholder groups represented on any board committees or advisory boards of the organisation	23
To what extent is your organisation investing in capacity building, including skills and technology transfer, in the communities in which you operate	45
Average	28

Table 6.17 Environmental integrity (illustrative scores)

Diagnostic questions	Score (%)
To what extent does your organisation engage in efforts to re-establish, protect or enhance biodiversity	24
To what extent has your organisation calculated the ecological footprint of all its operations (in terms of energy, carbon, water & waste)	20
To what extent has your organisation calculated the life cycle ecological footprint of all its products (in terms of energy, carbon, water & waste)	13
To what extent has your organisation achieved carbon neutrality	24
To what extent has total environmental impact (absolute volume of consumption, waste, emissions, etc.) gone down over the past 12 months	23
Average	21

6.3.3 Societal Contribution

Organisation X scores low overall in this category, with capacity building its strongest element. Practical next steps for improvement could include instituting stakeholder dialogues, cross sector partnerships, supplier audits and increasing charitable contributions (Table 6.16).

6.3.4 Environmental Integrity

The scale of ambition represented in this category probably explains why it is Organisation X's poorest performance area. Practical next steps for improvement include measuring total impacts, setting bold targets and implementing management systems to ensure impacts are reducing, and actively investing in biodiversity protection, and management of carbon, water and waste (Table 6.17).

6.4 Conclusion

Based on these illustrative self-assessment findings, Organisation X has demonstrated strengths in some areas—especially economic drivers and an orientation towards value creation—as well as revealing weaknesses that provide opportunities for improvement. The following recommendations could be made for moving forward:

1. In terms of the *Stages of CSR*, there are opportunities for improvement in all five stages, but especially in charitable CSR (notably employee volunteering and institutionalised giving). Organisation X should also map a CSR pathway to ensure that it is progressively maturing and moving towards the transformative agenda of CSR 2.0.

2. In terms of the *DNA of CSR 2.0*, Organisation X should build on its relative strength in value creation, while investing significant efforts in improving governance (especially reporting), societal contribution (especially supplier auditing) and environmental integrity (especially product life cycle footprint analysis).
3. In terms of the *Principles of CSR 2.0*, Organisation X should direct its creativity to social and environmental issues (especially supporting social entrepreneurs), while scaling solutions (especially using choice editing), responding to stakeholders (especially by supporting progressive policies), ensuring glocality (especially by adopting auditable codes) and designing for circularity (especially converting to renewables).

I trust the diagnostic, illustrative data and associated fictitious commentary gives some idea about the potential for further research on CSR 2.0 using this tool.

References

Hong H, Kostovetsky L (2012) Red and blue investing: values and finance. J Financ Econ 103(1):1–19

Grant AM (2012) Giving time, time after time: work design and sustained employee participation in corporate volunteering. Acad Manag Rev 37:589–615

Kronrod A, Grinstein A, Wathieu L (2012) Go green! should environmental messages be so assertive? J Mark 76(1):95–102

Koschate-Fischer N, Stefan IV, Hoyer WD (2012) Willingness to pay for cause-related marketing: the impact of donation amount and moderating effects. J Mark Res 49(6):910–927

Robinson SR, Irmak C, Jayachandran S (2012) Choice of cause in cause-related marketing. J Mark 76(4):126–139

Chapter 7
Conclusion

Abstract They key challenges to implementing CSR 2.0 are identified as complacency, admission and ambition. Five steps to begin practicing CSR 2.0 are described, including re-assess (performance), re-align (partnerships), re-define (purpose), re-design (products) and re-structure (policies). Finally, recommendations are made for fostering a more responsible form of capitalism.

Keywords Implementation · CSR 2.0 · Corporate social responsibility · Sustainable development · Leadership · Policy · Partnership · Impact assessment · Innovation · Future

CSR 2.0 is, at its core, about clarification and reorientation of the purpose of business. It is a misnomer to believe that the purpose of business is to be profitable, or to serve shareholders. These are simply means to an end. Ultimately, the purpose of business is to serve society, through the provision of safe, high quality products and services that enhance our wellbeing, without the erosion of our ecological and community life-support systems. The essence of CSR 2.0 is positive contribution to society—not as a marginal afterthought, but as a way of business.

7.1 Implementation of CSR 2.0

7.1.1 Challenges and Responses

The key challenges are:

Complacency—Many large companies have learned how to 'tick the boxes' of philanthropic, promotional or strategic approaches to CSR, resulting in complacency about their impacts and solutions.

Admission—This requires a genuine acknowledgement that business as it is practiced today is not sustainable and responsible, i.e. business still has a net negative impact on society and the environment.

W. Visser, *CSR 2.0*, SpringerBriefs in Organisational Studies,
DOI: 10.1007/978-3-642-40874-8_7,

Ambition—This requires that companies set audacious goals to reach sustainability, such as Walmart committing to zero waste and 100 % renewable energy. This is linked to the transformative CSR agenda.

To address these challenges, I propose:

Location—Classifying companies according to their stage of maturity on CSR—defensive, philanthropic, promotional, strategic or transformative. This will allow their leaders and the public to have an evolutionary path that does not end with adoption of CSR codes and standards, or GRI sustainability reporting (Algera and Lips-Wiersma 2012).

Education—Educating stakeholders to be more critical of the net impacts of companies, so that they become aware that current CSR practices do not adequately address the problem that originates with both our production and consumption processes and with shareholder-driven capitalism (Mackey and Sisodia 2013).

Celebration—Celebrating the leaders that are setting BHAGs—big, hairy, audacious goals—like Paul Polman of Unilever, which plans to double the company's size while reducing its environmental footprint and improving the lives of a billion people (Jordan et al. 2013). Conversely, naming and shaming leaders that settle for unambitious social and environmental goals (Kociatkiewicz and Kostera 2012).

To take the first steps, we must understand how to overcome resistence to change. According to Richard Beckhard and David Gleicher's Formula for Change (Dannemiller and Jacobs 1992):

$D \times V \times F > R$,

where:

D = Dissatisfaction with how things are now;

V = Vision of what is possible;

F = First concrete steps that can be taken towards the vision; and

R = Resistance to change

In terms of CSR and sustainability, the weakest variable is D, i.e. dissatisfaction with the status quo. This requires connecting people to the impacts of their actions, using tools like footprinting (for ecological, carbon and water), life-cycle assessment and supply chain auditing.

The evolution from CSR 1.0 to CSR 2.0 will be a long and bloody battle. In my own experience, as a South African, it took over 40 years of sustained and organised protest to change the entrenched power of the apartheid government, especially given the vested interests of big companies. The Occupy movement is one important indicator that the element (D) is changing, in a way that is favourable for CSR 2.0, and ought to be encouraged and sustained.

7.1.2 Taking Action

I recommend five actions to organisations that genuinely want to move from CSR 1.0 to CSR 2.0:

(1) *Re-assess*—This is about taking stock of the social, environmental and ethical impacts of company, i.e. creating a sustainability and responsibility performance baseline. Sustainability guidelines by the Global Reporting Initiative (GRI), Carbon Disclosure Project (CDP) and the International Financial Corporation's (IFC) are a good place to start, although ultimately this should embrace life cycle impact assessment (which resulted in BASF switching to recyclable Nylon 6), and full cost accounting (used at companies like Ontario Hydro and Baxter International).

(2) *Re-align*—This is about rethinking what cross-sector partnerships will shift perceptions and practices (Findlay-Brooks et al. 2009). In line with the eighth Millennium Development Goal (MDG), organisations need to find partners in business, government and civil society that will complement internal capabilities, while challenging the status quo. Examples include Rio Tinto partnering with the World Conservation Union (IUCN) to address biodiversity impacts and Unilever partnering with UNICEF, Oxfam, PSI, Save the Children and the World Food Programme to help improve the lives of more than a billion people worldwide.

(3) *Re-define*—This is about bold leadership, in particular setting a vision and strategic goals for the organisation, which will inspire and challenge all stakeholders (Visser and Courtice 2011). Examples include STMicroelectronics' Carbon Neutral strategy set by Pasquale Pistorio (Business Week 2005), Walmart's three sustainability goals for the company, including 100 % renewable energy, zero waste and making all products sustainable, defined under Lee Scott (Gunther 2006), Interface FLOR's Mission Zero set by Ray Anderson (Posner 2009) and Unilever's Sustainable Living Plan introduced by Paul Polman (2009).

(4) *Re-design*—This is about innovation, especially redesigning products and services to have minimal negative impact. Some companies are taking inspiration from Benyus's (2009) concept of biomimicry (learning from nature's designs), such as Interface FLOR's "gekko foot" non-glue adhesive tiles, while others are challenged by Prahalad and Hart's (2002) concept of the bottom of the pyramid (serving the poor), such as BP's Oorja low-smoke stove, or Vodafone's M-Pesa financial texting service.

(5) *Re-structure*—This is about transformation of the context in which organisations operate, i.e. changing the rules of the game, especially the policy environment. For example, supporting bold climate change policies to ensure a carbon price and efficient carbon trading (Fowlie et al. 2012), as the CEOs of the EU Corporate Leaders Group on Climate Change have done (Adey and

Visser 2007). Other examples include 'trading waste' through product take back schemes as happens in the electronics industry (Kellenberg 2012), and ingredient labels for salt, fat and sugar levels in so-called 'vice foods' (Mishra and Mishra 2011).

7.2 Towards Responsible Capitalism

At the heart of the transformational nature of CSR 2.0 is the need to embrace what McKinsey's Managing Director, Dominic Barton (2011), calls 'capitalism for the long term' and I call 'responsible capitalism'. This means testing all economic activity against five principles:

(1) *Investment*—ensuring that money is channelled towards productive investments and not into speculative trading in the casino economy, i.e. benefiting 'Main Street' and not only Wall Street (Lamin and Zaheer 2012), as the Co-operative Bank and Triodos Bank have demonstrated successfully.
(2) *Long termism*—understanding that real wealth is created by taking a long term perspective (Wang and Pransal 2012), as Generation Investment and Warren Buffet's Berkshire Hathaway practice.
(3) *Transparency*—embracing transparency in revenues, in line with the Global Reporting Initiative, Carbon Disclosure Project and Extractive Industries Transparency Initiative (EITI) (Wilkin 2009).
(4) *Full cost accounting*—internalising social and environmental costs (externalities), either through taxes (e.g. on carbon and pollution) and social and environmental profit and loss accounts (Gray 2010).
(5) *Inclusion*—enacting Porter and Kramer's (2011) concept of creating shared value, and serving the bottom of the pyramid (BOP) markets (Martin and Hill 2012), as demonstrated by the BOP 2.0 Protocol (London and Hart 2010).

We live in exciting times—a true period of bifurcation. We live on the cusp of the post-industrial revolution, and for the first time, we can finally glimpse what a new model of sustainable business and purpose-inspired capitalism could look like. As with so many things in life, the quest for a sustainable future is like a wheelbarrow. The only way we will make progress is if we pick it up and push forward. And the only way we will motivate people to join us in this effort is if they believe in what we are building. And what are we building? We are building nothing less than a new form of capitalism—one that serves society and sustains the planet.

References

Adey M, Visser W (2007) A new model of business-government policy dialogue on sustainability: the case of the corporate leaders group on climate change. Cambridge Programme for Industry Research Paper Series 3, University of Cambridge, Cambridge

Algera PM, Lips-Wiersma M (2012) Radical authentic leadership: co-creating the conditions under which all members of the organization can be authentic. Leadersh Quart 23(1):118–131

Barton D (2011) Capitalism for the long term. Harvard Bus Rev 89(3):84–91

Benyus J (2009) Biomimicry: innovation inspired by nature. HarperCollins, New York (in Press)

Business Week (2005) The race against climate change, 11 Dec

Dannemiller KD, Jacobs RW (1992) Changing the way organizations change: a revolution of common sense. J Appl Behav Sci 28(4):480–498

Findlay-Brooks R, Visser W, Wright T(2009) Cross-sector partnership as an approach to inclusive development. In: Utting P, Marques JC (eds) Corporate social responsibility and regulatory governance: towards inclusive development? URISD/Palgrave Macmillan, London

Fowlie M, Holland SP, Mansur ET (2012) What do emissions markets deliver and to whom?? evidence from southern California's NOx trading program. Am Econ Rev 102(2):965–993

Gray R (2010) Is accounting for sustainability actually accounting for sustainability… and how would we know? An exploration of narratives of organisations and the planet. Acc Organ Soc 35(1):47–62

Gunter M (2006) The green machine. Fortune magazine 154(3), 7 Aug 2006

Jordan J, Brown ME, Treviño LK, Finkelstein S (2013) Someone to look up to: executive–follower ethical reasoning and perceptions of ethical leadership. J Manage 39(3):660–683

Kellenberg D (2012) Trading wastes. J Environ Econ Manage 64(1):68–87

Kociatkiewicz J, Kostera M (2012) The good manager: an archetypical quest for morally sustainable leadership. Organ Stud 33(7):861–878

Lamin A, Zaheer S (2012) Wall Street versus. main street: firm strategies for defending legitimacy and their impact on different stakeholders. Organ Sci 23(1):47–66

London T, Hart SL (2010) Next generation strategies for the base of the pyramid: new approaches for building mutual value. FT Press, New Jersey (in Press)

Mackey J, Sisodia R (2013) The kind of capitalist you want to be. Harvard Bus Rev 91(1):34

Martin KD, Hill RP (2012) Life satisfaction, self-determination, and consumption adequacy at the bottom of the pyramid. J Consum Res 38(6):1155–1168

Mishra A, Mishra H (2011) The Influence of price discount versus bonus pack on the preference for virtue and vice foods. J Mark Res 48(1):196–206

Polman P (2009). McKinsey conversations with global leaders: Paul Polman of Unilever, McKinsey Q, Aug 2009

Porter ME, Kramer MR (2011) Creating shared value. Harvard Bus Rev 89(1–2):62–77

Posner BG (2009) One CEO's trip from dismissive to convinced. MIT Sloan Manag Rev Fall 51(1):6–7

Prahalad CK, Hart SL (2002) The fortune at the bottom of the pyramid. Strategy Bus 26:54–67

Visser W, Courtice P (2011) Sustainability Leadership: Linking Theory and Practice, SSRN Working Paper Series, 21 Oct 2011

Wang T, Bansal P (2012) Social responsibility in new ventures: profiting from a long-term orientation. Strateg Manag J 33(10):1135–1153

Wilkin S (2009) Heed the calls for transparency. Harvard Bus Rev 87(7/8):16–17

About the Author

Dr Wayne Visser is Director of the think tank Kaleidoscope Futures and Founder of CSR International. In addition, Wayne is Senior Associate at the University of Cambridge Programme for Sustainability Leadership and Visiting Professor of Corporate Responsibility at the Gordon Institute of Business Science in South Africa.

Wayne is the author of 17 previous books, including *The Quest for Sustainable Business* (2012), *The Age of Responsibility* (2011), *The World Guide to CSR* (2010) and *The A to Z of Corporate Social Responsibility* (2010). He has also published over 250 other works (chapters, articles, etc.) and has delivered more than 200 professional speeches all around the world, with his work taking him to over 65 countries in the last 20 years.

Wayne has been listed as one of the Top 100 Thought Leaders in Trustworthy Business Behavior (2013), Top 100 Global Sustain Ability Leaders (2012 and 2011), the Top 100 Thought Leaders in Europe and the Middle East (2011), the Top 100 CSR Leaders and the Top 20 Sustainability Leaders (2012 and 2009). He is also the recipient of the Global CSR Excellence and Leadership Award (2013), the Emerald Literati Outstanding Author Contribution Award (2011) and the Outstanding Teacher Award of the Warwick MBA (2010/11 and 2011/12).

Before getting his PhD in Corporate Social Responsibility (Nottingham University, UK), Wayne was Director of Sustainability Services for KPMG and Strategy Analyst for Cap Gemini in South Africa. His other qualifications include an MSc in Human Ecology (Edinburgh University, UK) and a Bachelor of Business Science with Honours in Marketing (Cape Town University, South Africa).

Wayne lives in London, UK, and enjoys art, nature, writing poetry and learning about new countries and cultures. A full biography and much of his writing and art is on www.waynevisser.com.

W. Visser, *CSR 2.0*, SpringerBriefs in Organisational Studies,
DOI: 10.1007/978-3-642-40874-8, © The Author(s) 2014

MIX
Papier aus verantwortungsvollen Quellen
Paper from responsible sources
FSC® C105338

If you have any concerns about our products,
you can contact us on
ProductSafety@springernature.com

In case Publisher is established outside the EU,
the EU authorized representative is:
Springer Nature Customer Service Center GmbH
Europaplatz 3, 69115 Heidelberg, Germany

Printed by Libri Plureos GmbH
in Hamburg, Germany